The Ultimate Engineer

Outward Odyssey
A People's History of Spaceflight

Series editor
Colin Burgess

The Ultimate Engineer

The Remarkable Life of NASA's
Visionary Leader George M. Low

Richard Jurek

Foreword by Gerald D. Griffin

UNIVERSITY OF NEBRASKA PRESS • LINCOLN

Library of Congress
Cataloging-in-Publication Data
Names: Jurek, Richard, author.
Title: The ultimate engineer: the
remarkable life of NASA's visionary
leader George M. Low / Richard Jurek;
foreword by Gerald D. Griffin.
Description: Lincoln: University of
Nebraska Press, [2019] | Series:
Outward odyssey. A people's
history of spaceflight | Includes
bibliographical references and index.
Identifiers: LCCN 2019005751
ISBN 9780803299559 (cloth: alk. paper)
ISBN 9781496218476 (epub)
ISBN 9781496218483 (mobi)
ISBN 9781496218490 (pdf)
Subjects: LCSH: Low, George M.,
1926–1984. | Aeronautical engineers—
United States—Biography. | United
States. National Aeronautics and
Space Administration—Officials and
employees—Biography. | Outer space—
Exploration—United States.
Classification: LCC TL789.85.L69 J87
2019 | DDC 629.4092 [B]—dc23
LC record available at
https://lccn.loc.gov/2019005751

To the Low family, with gratitude and admiration

Dr. George Low, the guiding light behind the Apollo program . . .
began his career as a research scientist and progressed to key
leadership positions in the manned spaceflight program. . . .
He continued his lifelong efforts to build a better tomorrow
while serving as president of Rensselaer Polytechnic Institute.
We're grateful for what George Low has done and the
ideas he stood for, and we'll miss him very much.
—President Ronald Reagan, 20 July 1984

The fundamental operating unit in George Low's NASA
experience has always been the team, whether he was a member,
contributor, or leader. But without him, those extraordinary
teams would have lacked the final sparks of ingenuity,
those extra quanta of strength that finally meant success.
—James C. Fletcher, NASA administrator

George Low knew as much about technology and about
humanity as any man I ever met. But what made him special
was that he knew how they fit together—whether in business or
education or in leading people in building a spaceship to the moon.
—Jack Welch, GE chairman and CEO

He was my favorite engineer.
—Neil A. Armstrong, NASA astronaut

Contents

Illustrations

Foreword

George M. Low was the deputy director of NASA's Johnson Space Center (JSC) when I joined the agency as a flight controller in 1964. He was many pay grades above me. In my eyes, he was a bigger-than-life leader in NASA management. I was familiar with the history of the creation of NASA and knew that George was a key player from the beginning. After the *Apollo 1* fire in 1967, I was selected to be a flight director in mission control, and I really got to work closely with George. More importantly, I got to know him on a personal level. George and I would become even closer later in our careers, but I could tell from those early interactions at JSC that he was a special talent and a special person. He was one of my all-time favorites—an inspirational leader, a terrific boss, and an overall great guy.

In many ways, George's style reminded me of another NASA legend—Neil Armstrong. Both were not big talkers, but whenever they spoke, everyone in the room got quiet and listened. George had a unique ability to get everyone's input, instantly synthesize and summarize all he heard, and then quickly make a decision. We trusted his decisions. George was a stabilizing force in almost everything we did, especially during the chaotic time right after the *Apollo 1* fire. His confident, calm leadership brought all elements of the Apollo Program together, and he helped us carry on. I can tell you for sure that everyone in the program wanted to do their best for George, because no one wanted to let him down. In my opinion, that is the hallmark of a truly great leader.

George was one of the hardest-working people I have ever known. I had the feeling that he probably didn't sleep very much. His days were very long, but he kept on going—always confident and inspiring. George was NASA's deputy administrator by the time the last Apollo mission was flown. He called me up to Washington to run NASA's Office of Legislative Affairs and to work with him on Capitol Hill. We were trying to keep the space shuttle on track and funded. George really was *the* go-to guy in Washington on the shuttle. We spent a lot of time on the Hill, explaining the shuttle, but we also covered topics that addressed NASA's programs in aeronautics, space science and technology, and research. George was on top of everything! When Apollo ended, we came pretty close to losing the shuttle and perhaps losing the entire manned spaceflight program. George played a huge role in saving the shuttle and getting us going again on all fronts. He didn't do it alone, but without George, NASA's future would have been bleak, at best.

George left NASA in 1976 to be the president of Rensselaer Polytechnic Institute, and it was about the same time I left NASA Headquarters. The next time I saw him, unfortunately, would be the last. I had returned to JSC as the center director, following Chris Kraft when he retired. George was then under treatment for a reoccurrence of melanoma at MD Anderson Cancer Center in Houston. My wife and I went there to visit him. I remember how very thin he was. But his spirts were high, and he greeted us warmly. He was determined not to let the cancer beat him, and his focus was firmly set on the future. I'll never forget what he said during our visit: "Gerry, I'll be flying on the shuttle one of these days." Unfortunately, he died shortly thereafter, leaving us at far too young an age. But in many ways, his spirit would go on to fly the shuttle—it was a program, like Apollo, that George helped bring about by his sheer force of will and dedicated hard work. And of course, his son G. David Low—who looked and talked much like his father—would go on to fly three successful shuttle flights. I am sure George flew in spirit on every one of those missions.

Too few people know or understand the important impact George Low had on the U.S. space program, let alone who he was as a person and the many great things that he accomplished during his life. He was never one to pursue the limelight, but he worked tirelessly and passionately behind the scenes to get us to the moon and back and to bridge the difficult transition of NASA in a post-Apollo world. I shudder to think what might have happened to NASA and our space program without him. It is my sincere hope that George's unique and remarkable story will inspire you to support this nation's continued efforts to lead the world in the peaceful exploration of space. It would be a fitting tribute to a man who dedicated his life to the same cause.

We need more George Lows in this world.

Gerald D. Griffin

Imagination has let us set our foot on the moon and our minds on the stars . . . gives us the hope for the future, a sense of direction, but it takes much more to reach our destination. It takes knowledge and courage and confidence—above all, the confidence to be willing to take risks. Yet today much of our society has an aversion to risk; people are afraid of science and technology, of where our experiments might lead us; afraid of the unknown and of the future. But if we are unwilling to venture, there can be no hope: without risk, there can be no progress.

—George M. Low, Inaugural Address, Rensselaer Polytechnic Institute, Troy, New York, 8 October 1976

Introduction

George M. Low once described himself as a dirty-hands engineer.[1] He said he wanted to be an engineer before he even knew what it meant to be one. In 1968, during the build-up to the historic *Apollo 8* circumlunar flight, Low explained what he meant by this description to World Book writer and space-age chronicler William Furlong. "I've always been interested in getting my hands dirty," said Low. "From as early as I can remember, I liked to fix things and take things apart. I like to get out on the floor and look at things for myself and find out how they work and why they're being put together that way."[2]

Furlong's syndicated profile of Low appeared in hundreds of newspapers along with a photograph of Low in his office at the Manned Space Craft Center in Houston, Texas. He is sitting at a long table filled with toggle switches and various machine parts from the Apollo spacecraft. Low kept these parts on hand to troubleshoot potential problems and design changes as he personally took on the critical task of redesigning the Apollo spacecraft after the deadly AS-204 (*Apollo 1*) fire in 1967. The fire cost the nation three precious lives and put the whole space program in doubt.

After the fire, NASA management called on Low to do the seemingly impossible. They asked him to bridge the anger, distrust, and animosity within Houston and bring the team back together to achieve their collective goal of landing a man on the moon. The secret of Apollo's success is paying attention to infinitesimal detail all the time, Low told *New York Times* science reporter Walter Sullivan.[3] With over 15 million individual piece parts in the spacecraft and rocket, it's not the big problems an engineer had to worry about. Those were obvious. Rather, it is the seemingly insignificant—a faulty toggle switch, damaged or frayed wiring, the wrong person in the wrong job. The work involved in safely getting a human being on the moon and back by the end of the decade would prove to be immense, and it would test Low and the entire NASA team to their limits. But in the end, they succeeded. And they succeeded in no small part because of George Low.

He often attributed his obsessive, hands-on work ethic as an engineer and a manager to his desire to know as much as possible about any given issue or situation. He abhorred the myopic limits of a single vantage point. As a result, he worked long hours. "I have always worked a little harder than most," he said. "I know some people who seem to have the utmost of self-confidence that whatever they start and whatever they do, they will do well. I've seen people like that fail, and fail very badly. I am not that sure of myself. I know my limitations."[4]

Despite the heavy, self-inflicted work load, Low never complained. He knew that lives depended on the thoroughness of the work. He believed that any failure was often predicated by the principal actors not asking the right questions at the right time. "I think the kind of question one has to ask, the motivation one has to have, and every decision one has to make," Low once said, "has to be based on the knowledge that one day you will have to tell the person who is going to fly the machine: 'I've done the best job I know how to do for you.' I put myself ahead—in time, not space—and say, 'If I make that decision in this way today, how will I feel at midnight in the Mission Control Center during the flight if something goes wrong?'"[5]

Low professed not to remember much about his formative years growing up in Austria.[6] It was a conscious decision on his part not to share those precious memories, even if he remembered more than he publicly revealed. He much preferred to talk about the present and the future rather than the past. "He was very much a visionary," said G. David Low, his son and space shuttle astronaut. "That was probably one of his best traits. The way I describe him is that he could see around corners. . . . I don't know anybody who can see around corners the way he could."[7]

Eva, his sister, recalls a memory of Low back in Austria, as an inquisitive young boy, spending countless hours playing with his Märklin erector set, building his own versions of the machines and equipment he saw on their father's farm.[8] There is a treasured family photo capturing such a moment. It shows Low as a dapper young boy, sitting at a table in the mid-1930s. He is patiently assembling what looks like a large motorized tractor wheel. His mouth is slightly agape in a satisfied half smile. He exhibits the same contented demeanor as he does in the World Book profile photo only thirty years later—a man completely absorbed by, and at peace in, the details of the work at hand.

That was George Low.

And this is his story.

1

New Beginnings

George Low (né Georg Löw) was born in Vienna, Austria, on 10 June 1926.[1] He came into the world during a violent, tumultuous time of geopolitical and social change—between two destructive world wars that would shape the ultimate destiny of his life in a myriad of ways. At the time, aviation was still in its early infancy, and modern rocketry was only a few months old. Global military conflict would propel the world into a new age of rocketry and aeronautics. It would unleash not only immense destruction but also rapid innovation and technological advancement. Indeed, the Wright brothers had flown at Kitty Hawk just twenty-three years prior, and already Lindbergh was planning for the first solo Atlantic crossing in 1927. Three months before Low's birth, Robert H. Goddard, a founding father of modern rocketry, had launched the world's first liquid-fueled rocket in Auburn, Massachusetts, on 16 March 1926. By the time Low himself would become a young aeronautical engineer a few short decades later, the aviation industry would be pushing the boundaries on altitude, speed, and distance with both planes and rockets, yearning to go to the planets and the stars beyond. Within the fullness of time, albeit half a world away, and in completely different circumstances from the life in which he was born, Low himself would become one of the world's greatest and most accomplished aeronautical engineers.

He spent his youth in Angern, Austria. A small, picturesque market town on the Morava River, it is located some twenty-five miles northeast of the capital, Vienna. Tall and thin for his age, the brown-haired, brown-eyed boy was known to be a private, introspective child. Whether riding a pony around the family farm, playing with his beloved toy train set, or building motorized machines with his metal erector set, Low was always intently focused on, and absolutely absorbed by, whatever it was that he was doing. Part of a wealthy and influential Austrian family of means and reputation, Low was also brought up in an environment that embraced education, innovation, and industry.[2] He learned English at an early age. Beginning in 1929 he had an English teacher one day a week. Over the course of the next eight years, he learned English through a form of language immersion in which the teacher talked to him in English, pointing out whatever it was she was describing. Increasing his English exposure, she also substituted for his governess whenever she was on vacation. A quick study, he advanced his language skills rapidly. He attended grade school in Vienna in the 1930s and then the Franz Joeseph Realgymnasium, a middle school, in 1936 and 1937.

During this time, the family owned significant real estate and agribusiness interests in and around Vienna. The brothers Wilhelm and Gustav Löw, his grandfather and great uncle, respectively, were born in the late 1800s in what is now recognized as the Czech Republic. At the turn of the twentieth century, they migrated from the Slovakian town of Vyškov to the eastern industrial region of Vienna in search of a better life. The brothers were successful hay and grain traders and instantly capitalized on the fertile farmland in and around Angern. In 1909 they built the largest and most modern distillery in the Habsburg Empire; by 1915 they expanded the distillery with a state-of-the-art cannery and refinery. Establishing Austria's only fertilizer factory shortly thereafter, they began recycling the plant and livestock biowaste from the farm. Soon, they began exporting fertile potash to as far away as Egypt, the Middle East, and the United States, as well as cultivating rare plant seeds. In just a few short decades, the Löw family had become a global agribusiness—one of the largest industrial concerns in Austria.

As part of their real estate holdings, they acquired Angern Castle, which became Low's childhood home. Along with its purchase, the family inherited the feudal title of *Grundherrn*—a medieval, landed estate title that granted them patronage rights over the area. By most accounts, the Löws took their responsibilities seriously and acted on them with humility and respect. They were especially generous in relation to the local Catholic church and Cloister Mauerbach—a monastery founded in 1313—as well as with the families that worked their farms and in their factories. The Löw family built and renovated apartments and houses for their employees. They also helped out socially. For example, whenever a family was expecting a baby, they would receive a *Kinderpaket*. This was a gift package filled with household necessities for raising the child. Every December, they hosted a Christmas party for the entire town.

All the children of Angern received a gift box filled with baked goods, fresh fruit, and a new pair of shoes.

As the family's success and influence grew, so, too, did its reputation and standing in Viennese society. Low's father, Arthur, and his aunt, Marianne, studied together at the University of Vienna. While studying, Arthur met Gertrude, Low's mother. All three eventually graduated with their doctorates: Arthur in chemistry, Marianne in economics, and Gertrude in liberal arts. Over the years, the family became friends and contemporaries with other famous Viennese, such as the Kinsky family, one of the noble Austro-Hungarian families dating back to 1237, and the Bloch-Bauers, another wealthy family with whom the Löws eventually partnered as fellow stock owners in the country's largest sugar factory. The Löws' influence also expanded to the realm of politics. It was not uncommon, for example, for the chancellor of Austria, Dr. Kurt von Schuschnigg, to be a dinner guest at their home.

Among the family's many economic interests, Arthur and Gertrude specifically focused their expertise on running the agribusiness. Professionally, they were interested in improving seed productivity and crop yields. They worked with scientists at the University of Vienna and elsewhere to develop better seeds, farming practices, and pest control. During visits to Angern Castle, many of the nation's leading agricultural scientists often joined the family for lunch and extensive conversations. Low and his siblings were often present at the table. During his formative years, he absorbed—if not directly, then indirectly—a lifelong passion for business and the sciences, at the feet of some of the best minds in all of Europe. He developed a passion for machines by being around the farm equipment, getting his hands dirty out in the fields. And he watched and learned as his mother took over the family business when Arthur died of cancer in 1934 at the young age of thirty-three—a precursor of the dramatic change his life would undertake during the coming years.

On the evening of 11 March 1938 Austrian chancellor Schuschnigg resigned under direct threat of invasion by Adolf Hitler, setting the dramatic stage for a bloodless coup and the *Anschluss*—the annexation of the country into Nazi Germany. Hitler, an Austrian by birth, wanted to unify the two Germanic nations. He also desperately coveted Austria's rich raw resources, skilled labor, industrial factories, and gold and foreign-currency reserves. The Löws' diversified industrial holdings and extensive wealth ranked them at the top of the Nazis' target list.

While the family had Jewish-blood lineage, they neither identified as Jewish nor practiced the Jewish religion.[3] Still, they were classified by the Nazi state as Jewish under the Nuremberg Laws. In Nuremberg, Germany, in 1935, the Nazi Party created a set of laws on citizenship and race. These laws dictated that anyone who had three or four grandparents of Jewish descent was determined to be a Jew, regardless of whether they practiced another religion or no religion at all. After the annexation of Austria by Hitler in early 1938, the Nazis imposed the Nuremberg Laws on Austrian citizens. Those classified as Jewish were denied full citizenship rights and were banned from certain professions, schools, and ownership of certain businesses—a development that had a significant impact on the Löws. It was the basis for their ultimate journey to America—a journey that would stretch out almost two years because of strict U.S. immigration quotas at the time.

The day after the Anschluss, a relentless campaign against Austrian Jews, Catholics, and proindependence supporters began. Within days, the Löws' businesses were seized by the Vermögensverkehrsstelle, a new government agency established by the Nazis to seize control of assets. On threat of imprisonment, Gertrude and the family were forbidden to enter or participate in the activities of their various businesses. While they were not explicitly confined to their homes, their activities were restricted. Although Low was allowed to attend school, he caught scarlet fever in April, which resulted in officials placing all of Angern under quarantine for six weeks—delaying the Vermögensverkehrsstelle from conducting an on-site inventory of the family's property and assets. When the quarantine was lifted, their personal assets were seized—including furniture, art works, and everything down to the silverware in the kitchen and most of their jewelry. They were evicted from Angern Castle and moved into an apartment in Vienna in the same building where Marianne, Low's aunt, was living. In due course, the family's personal and business assets were auctioned off by court order to compensate the Reich for millions in fictitious back taxes. By no sheer coincidence, the amount due by the family equaled, down to the pfennig, the total amount of their net worth.

To ensure cooperation, Nazi authorities confiscated the family's passports, which they were willing to ransom back and allow passage out of the country in exchange for the balance of their overseas assets. With conditions worsening by the day, especially for those classified under the Nuremberg Laws as Jewish, the family was left with little choice. They acquiesced, instructing their overseas bankers to transfer the money. Shortly thereafter, most of the family—including his grandfather and grandmother Wilhelm and Franziska Löw, his uncle Gustav Löw, and his aunt Marianne Hamburger-Löw—boarded a crowded train for Switzerland, but not before Gertrude extended help to other family associates, friends, and coworkers. Even after leaving Austria, Gertrude continued to help those she could in escaping Nazi-occupied Austria. She couldn't help everyone, but she helped as many as she could. As fate would have it, the family left barely one month before *Kristallnacht*—or "night of broken glass"—a violent uprising against Jews on the night of 9 November 1938. It is considered by many to be the beginning of the Holocaust.

Once out of Austria, they made their way to England after an eight-month stay in Switzerland.[4] In pre-Blitzkrieg London, they roomed for a short, three-month period in an apartment at Queens Gate Garden at about the time when World War II officially began in September 1939. As a result of the war, the family was evacuated to Dorset in late January 1940. During this time, Low attended several private boarding schools in both Switzerland and England. In Switzerland he attended Institut Montana Zugerberg, an international English-speaking school. In England he attended St. Paul's School in West Kensington before transferring to Claysmore in Dorset at the start of the war. Despite all the disruption, his report cards from the period show solid academic performance, a consistent ability to make friends, and an affinity for wood working and shop classes. He loved to work with his hands. The family's stay in both Switzerland and England were not without additional personal loss, however. Low's grandmother Fanny died in Switzerland, and Gustav would take ill, later dying in England.

Finally, in early 1940, as the war intensified and the conditions in England grew dire, the family's visas came through. They boarded the SS *Veendam* in Southampton on Friday, 26 January 1940, to begin their passage to America.[5] Also on board were grateful survivors of the Dutch SS *Simon Bolivar*— the first neutral ship sunk by German magnetic mines dropped into British coastal waters by the Luftwaffe.[6] The ship's sinking, just weeks before the *Veendam*'s departure, claimed 126 civilian lives. It was undoubtedly the subject of many late-night cabin whispers and nervous conversations aboard the *Veen-*

dam as they made their way from port out to sea. Due to the growing threat of mines along the main shipping routes, the Holland America Line engaged two British minesweepers to escort the *Veendam* on its journey.[7] On the first couple of nights of their voyage, the ship had to drop anchor and remain completely dark and still until daylight. They were avoiding not only mines but also Luftwaffe air raids.

While their voyage may have been filled with a number of grim threats of the war, the conditions they left behind were no better.[8] In wartime England, everyday essentials like food and clothing were rationed. Gas masks were standard issue and served as daily reminders of the horrors and destruction that German v-2 rockets would rain from the sky. In stark contrast, life aboard the *Veendam* provided a welcomed respite. Its ample amenities were more befitting vacationers and leisure travelers than its current cadre of refugees and immigrants fleeing the war. A double-stacked, four-deck passenger liner, the ship was outfitted with luxury staterooms and broad promenade decks.[9] Food was plentiful. The onboard cinema even featured the recently released *The Wizard of Oz*.

One can only imagine the impression this film might have made on a young Low, as well as on the other war-shocked passengers, especially upon hearing Judy Garland singing "Somewhere over the Rainbow." It is not hard to imagine Low standing out on one of the ship's long promenade decks at night, gazing up at the stars, perhaps wondering what his future in this uncertain world might hold. Nights out at sea would have revealed a star field to Low like none he had ever seen before. What awaited him on the other side of this dark, churning sea? The sounds of the ship's motors and the slapping of the water against the hull would have coursed through his ears.

After nine days of travel, the *Veendam* slowly entered New York Harbor for the Holland America Line pier in Hoboken, New Jersey, on the evening of Monday, 5 February 1940.[10] Low stood excitedly on the deck with his mother; his older sister, Eva; and his younger brother, Stefan.[11] The night air was chilly.[12] Snow and freezing rain threatened. Despite the inclement weather, they could not help but feel a growing sense of elation at the welcomed sight of the warm, glowing lights of New York City, starboard, and of Ellis Island and the Statue of Liberty, portside. The reassuring city lights of New York City ignited a well of emotion among the children as they slowly navigated the harbor. Pointing toward their destination from the ship's deck, the children began to talk excitedly in their native German. Aware of the momentous nature of their journey and the new life they were about to begin, Gertrude pulled her gaze away from the city. "Remember," she told them. "From now on we speak only English. From this point forward we will be Americans."[13]

While life in America promised to be a very different existence than the one they enjoyed in prewar Austria, there was reason for optimism and excitement. Like so many fleeing Europe, the Lows were grateful for simple survival. As fate would have it, and as the situation would require, Gertrude was not the kind of person who wasted her time talking about the past. Her steady, reliable nature and practical, stoic attitude clearly had an impact on Low—probably one of the more important and profound early influences on his own character. "All of us were expected to deal with what we had to deal with, and to do it in the best way we thought," said Eva, reflecting on the time and on his calm, pragmatic nature, a seemingly pervasive family trait. "We never discussed what might have been."[14]

***

Gertrude's older sister, Elisabeth, was already living in New York, having left Austria years earlier, and served as the family's visa sponsor.[15] She was married to a successful American cotton broker and inventor named Dr. Paul Monath. They lived in Hewlett, an affluent town on the south shore of Long Island. Many of the homes in this area of Nassau County consisted of mansions and harbor villas, some of which dated back as far as the American Revolution. In the excitement to enter their new country, a mistake was made on Low's immigration visa, listing him as "Georg Wilhelm Low," confusing his middle name, Michael, with the first name of his grandfather.[16] Not wanting to cause problems at such a critical time or, worse, risk denial of entry, Low signed his name as Wilhelm.

After a short stay with Elisabeth, Low and his family moved into an apartment on Long Island. Later, they moved to Queens. In the summer of 1941 Gertrude drove them west on a cross-country road trip to experience the expanse and natural beauty of the United States.[17] The scenic, leisurely trip was worlds apart from the nightly bombing raids and destructive battles taking place in Europe. There isn't much record of the details concerning their trip, but it seemingly helped to spark Low's lifelong passion for cars and long family road trips. In fact, he would emulate this car journey years later with his own children. During such trips, he would give his children tiny assignments, to quell the inevitable noise and chaos of five kids in the car at any given time.[18] They played word games and did mental math exercises that were cognitive development rituals. In some ways, it was very reminiscent of how he first learned English. These mental exercises were a game but a game with a purpose; the kids competed to improve their speed of response. Sometimes, to amplify the difficulty and fun, he had them unscramble anagrams.[19] In one instance, he challenged them with the phrase "tiny tiger" until they came up with the word "integrity." One imagines a similar learning experience as he traveled with his mother and siblings through the American countryside for the first time—playing English word games while taking in the sights, sounds, and experiences of 1940s America.

After that first road trip, the family settled into their life in New York. Eva would go on to earn her doctorate in chemistry from Yale in 1947. Brother Stefan, now going by the Americanized Stephen, would partner with Gertrude in running a successful, award-winning dairy farm in Milford, New York, before following his passion and establishing a regional commuter airline. Low attended and graduated with honors from the academically recognized Forest Hills High School in 1943.[20] Newly built in 1941, Forest Hills was often ranked as one of the best high schools in the country, especially in the arts and sciences. The school's sprawling, Georgian-style campus had a commanding view over the Flushing Meadows Corona Park, site of the 1939–40 New York World's Fair. The school was built with ten science classrooms and lecture halls, and upon its dedication, it was called "the most beautiful educational structure" in New York. Early Forest Hills alumni, along with Low, include Burt Bacharach, Art Buchwald, Ron Chernow, Art Garfunkel, Susan Isaacs, and Paul Simon.

At Forest Hills, Low flourished in math, science, and the language arts. After graduation, he enrolled at Rensselaer Polytechnic Institute (RPI) in the city of Troy, New York, where his career as an engineer would formally take root. He took the first step in correcting the mistake with his middle name by enrolling as "George Michael Low." Nestled comfortably on a rolling hillside overlooking the Hudson River, RPI is a private science and engineering research institute founded in 1824. It was the first school of science and civil engineering

to be established in the United States. RPI played a major role in the education of those who built America's industrial era infrastructure, and its students were known for being "builders of bridges." Now it was to play a major
part in the engineering education of a new bridge builder, one connecting
masses separated not by water but by the infinite vacuum of space. RPI was a
natural choice for Low, who finally learned what it meant to be an engineer,
among its faculty and classes. Years later he would return to lead the university in his post-NASA career. "Whatever I accomplished at NASA or contributed to the space program really began right here at RPI," Low explained in
his announcement as the institute's fourteenth president. "I owe my success
to where I got started. It may sound corny but that's it. I owe everything I've
been able to do to RPI."[21]

RPI had been recommended to him as a place to further develop his many
talents and interests, especially an affinity with mathematics. Once matricu-
lated, he took advantage of the wartime accelerated-learning program established to meet the nation's military need for engineers. In a year and a half, he
finished almost three years of academic studies. He also joined the Delta Phi
fraternity.[22] As an undergraduate, and later as a graduate student, his skills at
administration and his innate ability to engender trust among his peers saw
him serving in various leadership capacities at Delta Phi, including secretary,
treasurer, and, ultimately, fraternity president. During this time, Low lived
with his fellow fraternity brothers in a modest brownstone at 36 First Street,
in downtown Troy. Socially, he began dating a pretty young coed from Troy
named Mary Ruth MacNamara, or "Mary R." as she was known.

Initially intending to study mechanical engineering, he switched to aeronautical engineering at the beginning of his sophomore year.[23] His legendary
obsession with detail was already evident, as he felt mechanical engineering
was just too imprecise of a discipline. "You designed something in mechanical engineering, and then you just threw in a factor of 10 or 20 or 30," Low
said, somewhat tongue in cheek. He also found the relatively new subject of
aeronautics both fun and exciting. Who wants to worry about building the
same old Earth-bound bridges when you could be designing airfoils to influence drag and lift for faster and higher flight? This was, after all, the start of
the jet age, and aeronautics as a field was ascendant. In a personal letter in
1966 to a prospective RPI student, he explained that he had found a "spirit of

exploring a new frontier" within the aeronautical engineering field.[24] "When I got out of school, it was tremendously interesting, and incidentally, a great deal of fun," he wrote. "In looking back, however, I must say that much of the interest came from the fact that at the time aeronautics was a very challenging and fast-moving field. Both the jet airplane and the supersonic airplane were just coming into being; giant steps were needed and were made."

The first exposure Low had to the technical aspects of a jet was during a meeting of the Institute of Aeronautical Sciences in a lecture hall in the Ricketts Building.[25] The building, named after the university's eleventh president, was constructed in 1933 to house a newly established aeronautical and metallurgical engineering program. The department's wind tunnel and engine-testing lab were housed in the building's basement.[26] The tunnel was on a smaller scale to what he would later have at his disposal for experimentation in the massive wind tunnels at the Lewis Flight Propulsion Laboratory in Cleveland, Ohio; still, it was a runway for his future career path. The tunnel and the program were built by Dr. Paul E. Hemke, one of Low's earliest and most influential aeronautics professors. In a letter to Dr. Hemke shortly after the *Apollo 8* mission, Low mentioned the impact of Hemke's lectures on his understanding of the field.[27] "I still have with me the notes you used in my first aerodynamics course," he wrote. Those notes had also formed the foundation for Hemke's influential textbook, *Elementary Applied Aerodynamics*, which Low professed to still have "a well-worn copy of."

After this particular Institute of Aeronautical Sciences meeting, Low was hooked on aeronautics. His mind raced with the seemingly endless possibilities of the oncoming jet age, one in which he could dream of one day flying in such machines. In recalling the story, in March 1969, almost a quarter century later, he noted that it was common for him to fly over two hundred thousand miles a year in a jet.[28] What had once been a dream, just a few short years ago, had now become an everyday reality: "So, if you ask me if any of us have a chance to one day fly in space, the immediate answer has to be, 'No.' But, one day, it just might well happen."

Although he worked extremely hard during his first two years at RPI, Low claimed not to be a very good student,[29] a not uncommon feeling shared by many first-year engineering students. Even though he was at the top of his class in high school, he pointed out that *everyone* was a top student at RPI.

The competition was fierce. He entered the institute in the cold dead of winter, and he remembered having a mix of emotions that ranged from excitement at making new friends and learning new things, to dread and anxiety over not being able to keep up.[30] For the first time in his life, he was completely on his own. This realization crystalized for him in the fall of 1944 during a rare, momentary lapse of judgment. With his fraternity brothers, he tried to play an innocuous prank—putting RPI stickers all over a statue in a public park late one night. In a scene reminiscent of a Hollywood situational comedy, a local beat patrol officer happened to be standing nearby, almost instantly catching them in the act.[31] They were arrested and turned over to the university's president, and the charges were dropped. For Low, who was not yet even a citizen, the stakes could have meant more than just a reprimand or possible expulsion, which was worrisome enough. A blot on his record could have meant denial of citizenship. It was a wake-up call. "Sixteen was just too young of an age to go off to school," he said, recalling the story.[32] He needed to mature and grow up. Once again, that maturation and personal growth would come, in no small part, as the result of war.

Aeronautics, jets, RPI, and his romance with Mary R. would all have to wait. Just four months after he had registered for the draft in July 1944, his number had come due. An article in the 11 November 1944 issue of the *Troy Record* reports that Low was part of a U.S. Army preinduction group, drafted by Selective Service Board number 331.[33] On 13 December 1944 he and twelve other local young men assembled at the Palace Theater at eight in the morning and were transported to Fort Dix, New Jersey, where they were subsequently inducted into the U.S. Army.[34]

He shipped out to Camp Blanding, Florida, for three months of basic training. On 23 February 1945 he filled out a military petition for naturalization. He took great pride in letters back home to his family on being the first to complete the process.[35] It was a quick and relatively easy thing to do as a soldier about to be deployed overseas. During the process, he also formally Americanized his first name and corrected his middle name.[36] After becoming a citizen, Low spent the next six months in Corvallis, Oregon, attending the army's specialized training program where he studied topographical engineering.[37] He also earned his private pilot's license. In December 1945, near the end of the war, his orders came down for an overseas deployment to Frankfurt. "I went to Germany about the time they had 'Welcome Home' signs going up [in the states]," he said with his dry sense of humor.[38] Rather than dodging bombs and bullets, Low was stationed at the Army Corps of Engineers headquarters.

His job was to assist in the Army Corps of Engineers' massive, logistical infrastructure operations across Europe as it was pivoting from a war footing into a reconstruction force. In the corps, he gained exposure to government managerial and logistical expertise on a grand, unprecedented scale. During this time, the army's engineers had a wide range of duties, from logistical and support functions to building and maintaining barracks, hospitals, and airfields.[39] Additionally, the corps had to reestablish basic infrastructure such as roads, railways, canals, bridges, and ports—all of which lay in utter rubble and ruins.

Outside of just reading news reports or hearing stories told back in Troy, this was Low's first exposure to the fate of the German state that he had escaped only six years prior. There would not be much left of the cities and countryside that might have matched his memories. At the time of his deployment, the Austrian zone was under Russian control, and he could not visit.[40] The city of Vienna had been bombed over fifty-two times. The Soviet Vienna offensive in 1945 laid further siege to the city and its surroundings. Many of Vienna's finest buildings were severely damaged or destroyed. While he could not visit Vienna, let alone get anywhere near Angern, the conditions in Frankfurt and its refugee camps exposed him to an equally bleak and destructive tale. A month prior to Low's arrival in Frankfurt, former First Lady Eleanor Roosevelt visited Germany as part of a diplomatic tour. In her autobiography, *On My Own*, chronicling her life after the death of her husband, she describes the devastation of Frankfurt, Berlin, and Munich as Low would have experienced them at the time:

> I was stunned and appalled by what I saw as the pilot took our plane down to a low altitude and circled the ruins of Cologne and Frankfurt . . . places that I remembered as great and crowded cities. Later, when we circled Munich and then looked down on the rubble of Berlin, I felt that nobody could have imagined such horrible destruction. I knew that the Air Force deserved great praise for its role in winning the war, but I thought that nothing could better illustrate the sickening waste and destructiveness and futility of war than what I was seeing.[41]

The extent to which Germany and his former homeland of Austria had been damaged did not leave him unaffected. If he once allowed himself the futile exercise of imagining what his life might have been like had they not left Angern, all doubts now would have been erased. What might have been would have most certainly been horrific—a validation of his lifelong focus on the future, rather than the past.

***

After settling back into civilian life, he took up his aeronautical engineering studies at RPI with a sharper focus and a renewed vigor. He worked hard—in fact, harder than he had before. He finished under the GI Bill and graduated in January 1948 with a bachelor of aeronautical engineering degree. After graduation, he went to work for a short stint as an aerodynamicist for Convair in Fort Worth, Texas.[43] Convair had become a leading aircraft manufacturer on the cusp of expanding into rockets and missiles in the late 1950s, and the company's rapid expansion was spurred on as the military tried to bridge the so-called missile gap with the Soviets. In 1956 the U.S. Army's Air Material Command awarded Convair a study contract for a long-range ballistic missile that the company labeled HIROC, for high-altitude rocket. While the HIROC only made three partially successful test flights, the research it generated would eventually lead Convair to manufacturing the Atlas intercontinental ballistic missile (ICBM) rockets. These would be the same rockets that Low and his fellow engineers at NASA would adapt for use in human spaceflight during Project Mercury. As it was now, the practical experience at Convair led Low to want to go back to get more education.

After a few months in Texas, he reenrolled at RPI for graduate studies. Outside of improving his future career prospects, he also wanted to be closer to Mary R. In the preceding years, their bond had deepened, and they contemplated marriage. But first, he needed to secure his career. He needed an edge over the postwar glut of military-trained engineers now flooding the private sector, all competing for the same jobs. Motivated to get married and launch his career, he completed his studies in a year.

With competition for jobs being so intense, he decided to go into government work. His target was to pursue a position at the National Advisory Committee on Aeronautics (referred to as the NACA), the predecessor agency to NASA. New areas of experimental aeronautical research were taking place at the NACA. They were hungry for eager, talented, and creative engineers. In his autobiography, *Always Another Dawn: The Story of a Rocket Test Pilot*, famed experimental test pilot and engineer Scott Crossfield frames out why engineers of Low and Crossfield's caliber gravitated toward the agency in 1949:

> Unknown to the general public, the NACA had for years been the vital cauldron in which new ideas in aeronautical engineering were brewed and sampled. The agency was founded in 1915 by President Wilson, after the U.S. had lagged considerably behind Europe in the exploitation of the airplane for civilian and military purposes. The members of the committee, then the grandees of the U.S. aviation world, were charged with keeping close tabs on all domestic and foreign aviation developments, and to serve as a kind of clearing house for U.S. engineers.[43]

Low first applied to the NACA's facilities at Langley in nearby Virginia, without any luck.[44] He also applied to Ames, in California, because he had heard they were looking specifically for engineers from northeastern schools. Ames sent him a vague but promissory offer. With a job prospect in hand—no matter how vague—Low proposed to Mary R. She accepted, and they married on 3 September 1949 at Sacred Heart Catholic Church in Troy.[45] The newlyweds planned an extended, cross-country drive for their honeymoon, eventually ending up in California near Ames. At least, that was their plan, until a professor at RPI suggested that Low also apply to the NACA's Lewis Flight Propulsion Laboratory in Cleveland, Ohio. The facilities were newer. The lab was conducting research in cutting-edge hypersonic flight dynamics. There were a number of RPI graduates already installed at the facility, which would help his chances. Plus, if he got the job at Lewis, he and Mary R. could be closer to family in New York. Low had an old Delta Phi fraternity brother, Edgar M. Cortright Jr., who worked at the facility. Cortright and Low would go on to work together not only at Lewis but also throughout both of their future NASA careers. Low's application landed on the desk of Dr. John Evvard, Cortright's current boss and the chief of the Supersonic Propulsion Division. Evvard had two resumes for just one open position. On paper, Low appeared to be the superior candidate.

Nonetheless, Evvard asked Cortright for his opinion.

"Where do you think a guy like George Low might go in this agency?" Evvard asked.[46]

"I don't know," Cortright said. "But odds are, I will probably be working for him someday."

After a perfunctory interview, Evvard offered Low a research scientist position on the spot for the princely sum of a little over $147, every two weeks.[47] Low accepted without hesitation, sending his regrets to Ames. "He was a creative person," Evvard said, when reflecting on his initial impressions of Low. "He showed that spark of creativity. He was a deep thinker."

Abe Silverstein, Evvard's boss, was the newly appointed head of propulsion research at Lewis.[48] A founding father of what would become America's new space program, Silverstein was, at the time, the number two man in charge. At forty-one years of age, he was already a legendary, pioneering force in aeronautical engineering, especially to the young twenty-something engineers like Low. Silverstein's earlier work at the NACA facility at Langley resulted in critical-speed improvements of twenty-five miles per hour in World War II aircraft. He also was responsible for designing, building, and managing the facility's supersonic wind tunnels.

At Lewis, Silverstein liked to keep his core team small, and he surrounded himself with handpicked, smart people whom he respected.[49] Low became one of those people. Working within Silverstein's orbit would prove to be a critical professional relationship. "Abe was a very tough leader," Low recalled, naming him as perhaps the toughest but best boss he ever worked for at the NACA or NASA.[50] "I absolutely admired the guy. I worked closely with him, and I got to respect many of his qualities. What I learned from him more than anything was that to do a job, and to do it well, you have to dig in to the engineering and pay very close attention to the details."

Paying close attention to detail would become a personal mantra and foundational engineering philosophy for Low, especially after the tragedy of the *Apollo 1* fire in 1967. Attention to detail was the prerequisite for asking the right question to get at the right answers. "Once a question is asked, once a problem is properly identified and clearly defined, then a solution can always be found," Low said.[51] "In space, for example, we are successful in our efforts because we are curious, because we look for answers. And whenever we did have a failure, the reason was always the same. We had failed to be inquisitive. We had failed to ask the right questions. I know of too many people who have failed because they thought they knew everything and refused to ask questions or listen to others."

Bob Blue, a fellow NACA researcher and the man who showed Low around the office on his first day, underscored the importance of Low's mentorship under Silverstein. Blue and Low would become close friends during their NACA years. "Abe was a facilities man," explained Blue.[52] "He knew how to manage contractors and work with industry to get what he needed. As he assessed the needs of the airplane designers, he became equally interested in the aerodynamics problems and the mathematics of flight. But clearly his ability to successfully manage large-scale projects—like the design and building of the lab's eight-by-six and ten-by-ten wind tunnels —made him primarily a project manager." Blue credits this direct tutelage under Silverstein for Low's skills and abilities as a manager both of highly technical, skilled people and of large, diverse government projects.

"He learned a lot from Abe," Blue said.

During his tenure at the NACA, Low concentrated his research on experimental and theoretical areas of aeronautics such as heat transfer, boundary layer flows, supersonic turbulence, free-flight testing technics, and internal aerodynamics.[53] He would author, coauthor, and take part in the research for more than a dozen highly regarded research reports and technical briefs. These subjects were becoming increasingly important for achieving supersonic flight, that is, flight beyond the speed of sound. It was a time in aviation when engineers the world over were pushing the limits on speed, altitude, and performance innovations, all seeking air superiority in a still-turbulent and war-prone world. And Low's research was helping to rapidly advance the field. "When George and I worked together, we were research scientists trying to understand aerodynamics in supersonic flow," Blue explained.[54] "George was very heavy in mathematics, and he was applying mathematical theory and aerodynamic theory to results that he would observe in wind tunnel experiments."

At Lewis, Silverstein instituted a training program of classes focused on the cutting-edge topics they were researching.[55] He would assign different topics to the engineers within the branch to study and to teach. The program was designed to scale up their research and learning, and it proved to be invaluable in terms of keeping people on their toes and up to date on the latest trends. (The only other option staff had for continuing education was to take courses at Cleveland's Case Institute of Technology, but the fees were cost prohibitive for some of the young engineers.) Low helped to design the course offerings in that first year, as well as to teach the classes. He taught advanced aerodynamics theory and applied geometry as part of the four courses offered. "The classes were quite popular around the laboratory," Evvard recalled.

"We had an interesting mix of people," confirmed Blue, recalling the collaborative environment that Silverstein fostered.[56] "Everyone exchanged ideas daily. It was a team effort. We showed each other our work, asked for opinions and ideas, and gave credit to each other." And while they worked collegially in their research and day-to-day activities, the confidential nature and the importance of their work to national security and the military were never far from their minds. "Everyone in the lab had top secret clearance," explained Blue, and Low would maintain his throughout his career. "For example, one of our jobs was to study and review the highly confidential reports from Edwards Air Force Base on the progress of experimental flight-testing planes like the Bell-X1 (rocket) and the Douglass D558 Skystreak (jet)." Both planes were experiencing early developmental growing pains, especially with flight stability and turbulence. The engineers at Lewis were working hard to help find a solution.

The intense aerial matchup between the American F-86 Sabre jet and the Russian-built MIG-15 during the Korean War in the 1950s was also a subject of study and scrutiny. Both fighter jets employed innovative swept-wing designs that gave them marked speed and performance superiority over straight-wing jets of the era. While swept-wing designs have their roots in German military research dating back to 1935, the near-simultaneous appearance in the Korean theater of operations of these two similarly designed fighter jets raised deep suspicions within the U.S. intelligence community.[57] It turns out, it was no coincidence. Such swept-wing designs were being tested in the wind tunnels at Lewis. "The news at the time was filled with Senator Joe McCarthy's hunts for Communist spies," Blue said about the atmosphere around the lab. "When the news broke about Julius Rosenberg in the summer of 1950, it didn't mean anything to us at Lewis until the FBI showed up and started asking questions."[58]

Julius and Ethel Rosenberg were accused of giving the Soviet Union top secret military and atomic bomb technology, as well as classified radar, sonar, and jet propulsion secrets. They were convicted in 1951 of espionage and executed in 1953. In March 1951 the FBI announced the arrest and indictment of thirty-two-year-old William Perl, the former head of the Special Projects Branch in Cleveland.[59] Perl, Low's boss at the time, was initially considered a suspect in leaking top secret Lewis plans to the Russians; he would go on to be charged and convicted of lying in connection with the Rosenberg spy case. According to Blue, Perl's arrest was a shock for everyone at Lewis.[60] For Low, the Perl case would set into motion a series of rapid management changes that would soon see him promoted from individual researcher into increasing levels of managerial responsibility. Recognized by Silverstein and other leaders at Lewis for his natural ability to manage and motivate people, as well as his creative intellect and meticulous attention to detail, Low advanced rapidly. In short order, Low became head of the Fluid Mechanics Division from 1954 to 1956 and then chief of the Special Projects Branch from 1956 to 1958.

Given the focus on X-15 rocket plane research, Neil Armstrong spent five months as a test pilot at the NACA Lewis in 1955, during the middle period of Low's tenure at the facility.[61] Like Low, Armstrong was studying high-Mach heat-transfer issues. While they didn't get to know each other very well at the time, their paths did occasionally cross.[62] "The only product of the NACA was research reports and papers," Armstrong said, describing the environment in which they both worked at Lewis.[63] "When you prepared something for publication, you had to face the technical and grammatical 'Inquisition.' The system was so precise, so demanding... they went into that kind of detail." It was an environment that would propel Low's meticulous attention to detail into overdrive, striving for a precision and exactness not only in research results but also in the reports that were issued. As an artifact of learning English as a second language, his attention to word choice and grammatical style was a perfect fit for managing the work product of the lab.

As a result, employees at the NACA or NASA could not escape Low's editorial pen, especially those who used words as their daily tools of trade, such as in the public affairs department or in the daily production of the flood of bureaucratic memos that served as any government agency's lifeblood. While others might write their editorial comments in common black or blue ink, Low decided early in his career to use a green felt-tip marker. In doing so, he knew his comments would stand out, and the recipient would know instantly that it was coming from him. Known as his "green hornets" or "green stingers," Low's green felt-tip pens would become his signature hallmark. Much to the chagrin of his children, he was a strict grammarian at home too.[64] He would correct their school papers and homework with the same critical eye.

The Low household was always well stocked with green felt-tip markers.

By all accounts, Low was respected as a manager by his NACA direct reports and colleagues. In addition to being comfortable with technical issues, he was proficient at formulating and dealing with lab policy and administrative matters. He was also particularly strong in developing and inspiring people. While he could be as tough on his staff as Silverstein, he asked nothing of them that he wouldn't expect of himself. With Low, criticism wasn't personal. The only goal was good work. Those who worked with and for him described him as a confidence-inspiring manager who never seemed to lose his cool or his perspective.[65] In fact, no one could recall him ever raising his voice in anger or getting overly emotional. "He didn't shout or scream, ever," confirmed Dorothy Reynolds, who worked with Low when he was president at Rensselaer.[66] She knew and worked with him for what was then most of her adult life. As he had for many other people in whom he recognized great potential, he became a mentor to her and encouraged her to get her doctorate and to go on to bigger and better career opportunities. "In fact, when he was really upset, instead of getting louder, his voice would get lower," she recalled. "He would get very articulate, and his comments became like a surgeon's scalpel." Those on the receiving end never failed to recognize that they had crossed a line. Very few dared to repeat the error. There was never mistaking Low's intent. "Oh, they knew. They knew," she said, delighting in the memory when his calmness would put rude, know-it-all people in their place.

During the Cleveland years, Low and Mary R. immersed themselves into postwar suburban life. They built a house out in Middleburg Heights, about two miles from the office.[67] They welcomed the first four of their eventual five children: Mark, Diane, George David, and John. Nancy, their youngest, would be born later, after the family moved east. During the work day, Low managed his teams and worked on his research projects, and he taught his noncredit graduate courses. At night and on weekends, he and Mary R. spent time with the children, as well as with their close circle of NACA friends, like the Blues and the Cortrights. Family was always important to him, and he remained fiercely loyal to them. They liked going dancing and listening to music. Although he enjoyed most forms of music, he was particularly fond of jazz.[68] He amassed a large, eclectic record collection of artists from Dizzy Gillespie to Herb Alpert. In relaxed moments, he could be caught humming or singing to his favorite tunes. In his later post-NASA years, his family remembers him humming songs from Sinatra's *September of My Years* album, especially the hit "It Was a Very Good Year."

In the summer months, the NACA hosted many large, family-oriented picnics. Occasionally, they gathered for group dinners as well. As a tight-knit community, they worked on each other's home-repair projects.[69] They'd take turns building fences, painting garages, putting in lawns, and, of course, installing kitchens and appliances. True to his nature, Low reveled in being a hands-on engineer, at home as well as at the office. At the office, he liked to work in the experimentation cells and with the test models, exploring the true engineering side of the business more than just the purely theoretical research. At home, he would often take it upon himself to fix a broken appliance or rewire a light fixture. In fact, his children do not remember a time when he needed or, for that matter, wanted to call a repairman. He once spent an entire afternoon taking apart a malfunctioning washing machine, spreading out the various wet parts to locate the one telltale and out-of-place element responsible for the malfunction. He also liked to buy old cars and work on them in the garage on Saturdays—a hobby that he maintained throughout his adult life. As for his garage, he built it himself. He even built a weekend getaway cottage on the shores of Atwood Lake, Ohio, with his fellow NACA engineer Ed Cortright, where the families would go for fishing and relaxing on the weekend.

In short, the Cleveland years were an idyllic family time, well balanced with important engineering work that he enjoyed. After the tumultuous first few years of his life, he was now on course for a quiet, predictable middle-class life for himself and his family in the sleepy, post–World War II suburbs of Cleveland.

Until, of course, the Soviet Union launched Sputnik.

Then everything changed.

2

A Man in Space, Soonest

On 4 October 1957 Low was driving south from Cleveland on old Route 21 with Mary R. and the kids.[1] It was a breezy, cool Friday evening. They were traveling to their cabin on Attwood Lake for a weekend of relaxation. The family listened to the radio as they passed through the bucolic small towns that dotted the route. It was already an eventful news day: racial tensions continued to boil over in Little Rock, Arkansas; a flu epidemic raged across the nation; the election of Jimmy Hoffa as head of the Teamsters was underway; and, on the lighter side, extensive discussions about the upcoming Milwaukee Braves and New York Yankees matchup in the third game of the 1957 World Series.[2]

At some time between seven and eight o'clock, toward the end of their journey, he heard the announcement that Russia had succeeded in putting into orbit the first artificial satellite. The world stopped and took notice. So, too, did Low. For some time, both the United States and Russia were looking at using intercontinental ballistic missiles (ICBMs) for launching orbital payloads. Some in government and the military—such as celebrity rocket scientist Werner von Braun in Huntsville, working at the time for the U.S. Army's ballistic missile program—foresaw the space age coming.[3] Like the proverbial hare racing against the tortoise, they warned the United States that it was falling behind. Their warnings were not taken seriously outside of their own circle of converts. For Low, as well as the rest of the stunned world, Sputnik marked a dramatic turning point. "It was clear to me we were beginning a new era. It was a momentous and important historic occasion. Sputnik was a messenger of change," Low said of the moment when he first heard of the Russian accomplishment.[4] "Sputnik's first signal told us that we in the United States no longer had a monopoly on advanced technology."

Within the hour, an RCA receiving station in Riverhead, New York, would pick up Sputnik's technopulse signal and rebroadcast it around the nation on NBC radio.[5] As the tiny, silver Russian satellite orbited the earth every ninety-eight minutes, it sent shockwaves from Moscow to Washington and along every rural byway in the United States with access to a radio. Until now, America's space program had been largely the uncoordinated and classified purview of various military programs. In *Chariots for Apollo: A History of Manned Lunar Spacecraft*, authors Brooks, Grimwood, and Swenson perfectly capture the impact of the Russian achievement as Low and his NACA contemporaries would have experienced it:

> The orbit of Sputnik I in October 1957 stirred the imagination and fears of the world as had no new demonstration of physics in action since the dropping of the atomic bomb. In the United States the effect was amplified by the realization that the first artificial satellite was Russian, not American. . . . Sputnik I caused alarm throughout the United States and the ensuing public clamor demanded a response to the challenge. During the next year, many persons in government, industry, and academic institutions studied means and presented proposals for a national space program beyond military needs. After decades of science fiction, man himself, as well as his imagination, moved toward an active role in space exploration.[6]

Hearing the messenger of change, Low was eager to answer the call. He had not grown up as some of the others, with his nose stuck in science fiction books like *From Earth to the Moon* or other imaginative yet loosely scientific accounts of spaceflight. For him, the excitement was about the new applications of technology and the creative engineering that was necessary to achieve such things. Those already involved in the U.S. space-exploration efforts would soon call on Low and his technical skills and creativity, putting him on the fast track in America's fledgling space program. Up until Sputnik, the Lewis Lab had very little to do with spaceflight or rockets. Its main research focus was on propulsion, and it had been working on transforming the turbojet—a bulky, roaring, fuel-thirsty engine—into a quiet, dependable, commercially viable system.[7] Rocket engines and motors? Not so much. And not very seriously. In most of the NACA circles prior to Sputnik, the field was still considered a juvenile pursuit of science fiction fanatics and Buck Rodgers wannabes. Low, a self-described and unabashed futurist, wasn't as dismissive. By this time, he had already been running calculations and working with Ernst Eckert on research concerning heat-transfer problems encountered in nuclear aircraft propulsion, and he was a supporter of the lab's early rocketry experiments.[8] As he would do throughout his life, Low set his gaze on the future, around the corner from where current thinking and the status quo currently resided. It was never really good enough for him to just tinker with the status quo; he liked to stretch the reach of capability, to creatively challenge conventional thinking.

After Sputnik, Hugh L. Dryden, director of the NACA, chaired a meeting of all the NACA laboratory directors and associate directors in December 1957 to discuss their future role in this new age. This resulted in each NACA laboratory proposing ideas for a space initiative. "Engineers have a special responsibility to be bold and imaginative," Dryden said in early 1958, setting out a long-range vision within the nation's nascent space program.[9] "The goal of the program should be the development of manned satellites and travel of man to the moon and nearby planets." It was a goal and a responsibility that Low completely embraced; he would gain the reputation over his career within NASA for making some of the boldest decisions in the space program.

At the time, there was some serious research being performed on the fringe at Lewis by a small, relatively underfunded staff. They were studying different rocket fuels, including liquid hydrogen—an area of niche interest to Silverstein, as it might also apply to their work with high-altitude aircraft such as Lockheed's U2 spy plane.[10] Liquid hydrogen, the critical future fuel source for the Saturn family of rockets that would eventually take the nation to the moon, as well as the fuel source for the space shuttle's main booster engine, would go on to become the American space program's fuel of choice in part because of its low molecular weight and high intensity burn rate. But it was a tough fuel to learn how to use, given the technical challenges created by the fuel's need to be kept at extremely low temperatures. Foundational research at Lewis during this time would prove crucial in NASA's ability to tame the fuel into a reliable, usable staple of U.S. spaceflight.

In the year prior to Sputnik, Low had put together a training seminar on hypersonic flight. It was called "From Mach 4 to Infinity." The seminar explored propulsion problems related to aerodynamic heating at high altitudes and speeds as they would be experienced by winged aircraft like the x-15 and other similar experimental craft flying at the edge of space. "The lab was mainly working on compressors and turbines and that sort of thing," Evvard explained.[11] "George knew that if we presented a course that was really far out, it might change the whole viewpoint of some of the laboratory personnel. It was really the start of the spaceflight effort at Lewis." While the NACA overall demurred in its focus and research on rockets and spaceflight, Silverstein was cognizant of this new technology's applications. He hoped the seminar series would inspire others at the laboratory to get into the new field. It worked. "It inspired a lot of people," Evvard confirmed. "The second year, people mobbed the course because those who took it the first year were raving about the insights that it gave. It initiated a lot of new thinking at the laboratory."

Given the growing demand, Silverstein expanded the rocketry staff and their budget.[12] He gave them an express charge to look at the potential of rocket propulsion over air-breathing engines. Low assigned one of his designers, Robert Godman, to develop with Robert W. Graham an operational wind tunnel cell for rocketry experiments. With engineers such as Warren North and Bruce Lundin, Silverstein's rocket team had just begun designing a hydrogen-powered second stage, which was to be launched on top of a U.S. Army Sergeant single-stage rocket. "We had it pretty well along in terms of dynamic stability and propulsion when Abe got the call [from Dryden] to go to Washington to help put together the space program," recalled North.[13]

Meanwhile, Robert Gilruth and his start-up Pilotless Aircraft Research Division (PARD) over at the NACA Langley in Virginia were conducting research with small missile launches off Wallops Island. The only other NACA facility that used Wallops for experimentation was Lewis. The teams were used to working together. At Wallops the PARD team tested the effects of high-speed flight on materials and aerodynamic designs. Specifically, they were evaluating the reentry heating problems of the rocket payloads for both ballistic capsules and missiles, learning quickly through experimentation that a blunt rather than a sharp nose cone was the best approach.[14] The PARD team and many of its members, like Christopher Kraft and Max Faget, would later become important and influential members along with Low on NASA's Space Task Group (STG).[15] As a team, the STG, managed by Gilruth, would eventually be tasked in operationally realizing America's manned spaceflight ambitions and would be the technical driving force behind much of the success in landing a man on the moon in 1969.

Despite the excitement generated after Sputnik, not everyone within the old NACA was on board. At Lewis, for example, the shift to more rocket and nuclear propulsion research, while prescient and in-line with Dryden's vision, created a bit of a rift in the NACA Lewis aeronautical research culture. There was some concern among NACA careerists that if they got into the spaceflight business, they would move out of the realm of theoretical research (their traditional main bread and butter) and become more of an operational arm of the agency.[16] They feared losing their purely academic culture. They wanted to hang on to the many areas of pure aeronautical research they were now pursuing. Unfortunately for this group, and fortunately for Low, it was becoming the minority opinion.

Nudged on by Sputnik, James Doolittle, chairman of the NACA's Main Committee, commissioned Dr. H. Guyford Stever of MIT to set up a special committee for the NACA to explore space technology, in November 1957.[17] Over the course of the next year, this committee studied human factors and capabilities in space exploration. Human factors would become an area of great interest for Low, especially when he considered the role and skill qualifications for astronaut selection as part of the future Mercury program. The Stever committee reported, "The ultimate and unique objective in the conquest of space is the early successful flight of man, with all his capabilities, into space and his safe return to Earth. Just as man has achieved an increasing control over his dynamic environment on Earth and in the atmosphere, he must now achieve the ability to live, to observe, and to work in the environment of space."[18]

Low embraced the Stever report's findings, but his ambitions for the future program would prove to be greater than just going into space and coming back.

In March 1958, amid the Stever committee's deliberations and the PARD experiments off Wallops, a debate raged in Washington about which existing government agency should take over coordination of the nation's space efforts.[19] Should it be the NACA, with its surfeit of experienced aeronautical engineers like Low and their expertise in applied research? Should it be the U.S. Army, including von Braun and his team of German rocket scientists out in Huntsville, working on the Redstone and dreaming up larger booster systems like the Saturn? Should it be the U.S. Air Force, under their Convair-built Atlas ICBM program? Or should it be the U.S. Navy, with its ambitious Vanguard rocket project?

There were many options.

Luckily for Low, the pendulum of fate swung the NACA's way.

President Eisenhower—not particularly a space fan, despite the obvious military applications that rocket technology presented—opted for the national space effort to be run and coordinated by a new, independent civilian agency in as open and transparent a way as possible. The country's new space agency would be built around the NACA's extensive infrastructure and experienced research capabilities. It would tap into and coordinate with the various military agencies and private industry as needed. Eisenhower signed the National Aeronautics and Space Act into law in July 1958. He appointed T. Keith Glennan, president of the Case Institute of Technology in Cleveland, as NASA's first administrator. Dryden, with the help of his former NACA leadership teams, including Silverstein, would be deputy administrator, overseeing NASA's technical and scientific work.[20]

As things were progressing quickly in Washington, Low and his fellow Lewis engineers back in Cleveland were supporting Silverstein with research and counsel.[21] "Abe had a *sub rosa* group of seven or eight people like George Low putting together a plan for a space agency out of the work we had done at Lewis," said Frank E. Rom, a NACA Lewis theoretician on nuclear propulsion for airplanes and interplanetary rockets.[22] It was Silverstein's own space applications kitchen cabinet, and Low was at the table. Specifically, Low and the team helped Silverstein to figure out the part that manned spaceflight—as opposed to unmanned exploration with satellites—would play in the new agency.

Similar to his early days in aeronautics, Low found the work fun and exciting: "I told Abe one Sunday if there is anything in the space business he could use me for, I'd sure like to join him."[23] Low, however, made sure to emphasize to Silverstein that he did not want to go to Washington just to get caught up in the bureaucracy of the place. He didn't want to get lost in some administrative position. Despite his strong theoretical research background, Low wanted to get his hands dirty in the trenches—he wanted to be on the operational side of the business, turning research into reality. "Goddard was going to be the spaceflight center, and it was already a gleam in Abe's eye at the time," recalled Low.[24] "He agreed with me, and he said he would put my name on the list of whatever might come up when he eventually formed Goddard." Low should have gotten that promise in writing.

***

While Low would not truly come into his full, historic stature within NASA and manned spaceflight until Apollo, his importance from the earliest days of NASA would prove to be nonetheless as impactful and important. For without his efforts, it is highly doubtful that Kennedy would have seized on the moon as a destination within the decade when he announced it in 1961. In October 1958, just one month before the NACA officially became NASA, Silverstein tapped Low to come to Washington for a couple of weeks to help out.[25] Low was just thirty-two years old, but Silverstein needed his technical creativity and skills as a writer and a deep thinker. Abe assigned him to work with Bob Gilruth and the team tasked with putting together the final plans for what would become Project Mercury.

A month earlier, Glennan and Roy Johnson, a former General Electric vice president and now the director of the Defense Department's Advanced Research Project Agency (ARPA), had agreed that their two agencies should join forces for a Man-in-Space-Soonest program.[26] ARPA had been formed in early 1958 to oversee the military's various and uncoordinated projects on space research. In a rare form of government interagency foresight, Glennan and Johnson wisely decided to work together to reduce potential budgetary, operational, and congressional friction. They formed a joint NASA-ARPA Manned Satellite Panel. The panel's mandate was to draw on the best "state-of-the-art" ideas that would get a U.S. manned mission into orbit the quickest and with the lowest cost possible.

The panel would be chaired by Gilruth, and it would include Max Faget from Langley. It would be the first time that Low would work directly with the visionary likes of Gilruth and Faget.[27] He would end up working closely with both throughout his career, and Gilruth would become a seminal mentor, colleague, and friend, on par with—if not exceeding—Silverstein. A modest, balding engineer from Minnesota, Gilruth was not a man who was prone to imposing his own ideas on a team.[28] Rather than micromanage, he had a habit of just asking his team members questions, which often nudged them along in the right direction in their research and work. Low would seize on this Socratic management style and use it to skillful effect for the rest of his career. For Apollo astronaut and fellow aeronautical engineer Ken Mattingly, Low's encyclopedic focus on and access to detail, as well as his Gilruth-like ability to ask the right questions, were equally as impressive and important. "I've seen a lot of good people, but I've never seen anybody of his caliber," Mattingly said. "George Low had this way.... He would just ask questions, based on his knowledge that there was more to a story than was perhaps immediately obvious."[29]

Low accepted Silverstein's offer. "George arrived as a pioneer in the office of spaceflight programs," Silverstein said.[30]

It was a unique chance to learn more about the country's various space initiatives, especially the booster programs being cooked up by the U.S. Army down in Huntsville under von Braun. From Gilruth's perspective, Low would become an indispensable and critical member of a small, impactful team. "He was good at everything," Gilruth recalled.[31] "He was a top-notch engineer and scientist. He could handle people well. He was good with money matters. And he was very easy to get along with. Yet he wouldn't stand for any monkey business. He was an ideal administrator and friend and a big help to us, especially in the days when we didn't have many people. He was worth about ten men."

"We had a task force putting together the final plans for Project Mercury, spending two weeks putting it all in a very simple document," Low said about the joint NASA-ARPA Manned Satellite Panel.[32] Despite a number of competing theories at the time, the group worked day and night, discussing various approaches to come up with a group consensus. Their eventual report, titled *Objectives and Basic Plan for the Manned Satellite Project*, outlined very quickly and eloquently in two and a half pages NASA's first major foray into manned spaceflight.[33] It was built off the already accumulated research and project planning from a myriad of various NACA, Department of Defense, and private-industry initiatives.

It would result in Project Mercury.

The experience gave Low the opportunity to review state-of-the-art space planning beyond just the work being conducted at NACA Lewis. He immersed himself in the philosophies and details of such dominant theories as the so-called von Braun paradigm and the advanced rocket booster research that was being conducted for the U.S. Army in Huntsville. In 1994, political scientist and space historian Dwayne A. Day coined the term "von Braun paradigm," which he boiled down to a linear and progressive strategy for interplanetary space travel as proposed by von Braun in his earlier work. Von Braun explained his strategy in a series of visionary articles in *Collier's* magazine between 1952 and 1954. In short, the von Braun paradigm starts with rapid, airplane-style access to space via a reusable space shuttle, progresses on to building a space station, and then extends out to trips to the moon and then on to Mars and beyond. It's a linear, iterative approach to eventual deep-space exploration, which Low endorsed.

In assessing the current state of planning and technological capabilities in 1958, Low was convinced that human versus machine space exploration held some of the most promise. "It is impossible for me to conceive an instrument that could effectively and reliably duplicate man's role as an explorer, a geologist, a surveyor, a photographer, a chemist, a biologist, or a host of other specialists whose talents would be needed in space exploration," he told science editors at a United Press International conference in 1960.[34] He said the goal of America's space program would be for the nation to first learn how humans react in space, what their capabilities were for performing work, and what must be provided in the vehicle to allow them to function usefully. "Equally important," Low underscored, "we should learn a good deal about the design, engineering, and operation of the future craft for man's flight into space."

He firmly believed that the human element—the astronaut—was destined to play a vital and direct role in the future pioneering flights. Humans, Low told the assembled crowd of science and aviation journalists, "are the most versatile scientific instruments yet devised." In the ensuing years, he would work very hard to ensure proper integration of human-factors engineering. He also was an early supporter of NASA's own space medical research in areas such as radiation exposure and the physical stresses of spaceflight. In fact, he believed that the biosciences—physiology and biology—would eventually become more important than actual technical aeronautical engineering. "I think you must consider the biosciences a more forward-looking field than the physical sciences," he explained. "Just think of the information-transfer that exists in the human body. We do not have a machine that can do anywhere near that yet."

Dr. George M. Knauf was the deputy director of aerospace medicine for NASA.[35] He would work closely with Low in those early, pioneering days. After World War II, Knauf had already been working within the Department of Defense medical services supporting the U.S. Navy and the U.S. Air Force, investigating biomedical aspects of exposure to microwave radiation via radar and other radio-frequency-emitting electronic equipment. NASA engaged Knauf for similar purposes, exploring issues of space radiation exposure and the biomedical stresses of spaceflight on humans. Many members of Congress, especially Congressman Emilio Daddario of Connecticut, objected to what they viewed as a duplication of efforts in the military and NASA on biomedical sciences work—just one of many overlaps in those early days. Over the years, Low would support Knauf's work in congressional testimony for continued funding and support. He viewed Knauf's work as vital to the future of human spaceflight, independent of military needs.

In a letter to Low in 1964, a grateful Knauf wrote,

> Your contribution in making sense, for the first time, out of NASA's space medicine program was indeed a significant one. But even more important, to my mind, is the relationship that you have been able to establish with the Department of Defense in this area. You must have been justly proud when as severe a critic as Daddario congratulated you after your Congressional testimony earlier this year. I am pleased that you saw fit to join NASA as a permanent member of the organization; and I am pleased to have a friend like you in the Office of Manned Space Flight.[36]

Low would also become critical in convincing Bob Gilruth of the limited dangers of radiation to the astronauts for future planned missions to the moon. "All during the early planning for the lunar missions, I had been greatly concerned about the effects of solar radiation on the astronauts," Gilruth recalled.[37] "Experts were not all in agreement as to the amount of radiation that might be received on a mission to the moon. I remember George Low stating that the normal shielding of the cabin walls, together with the low probability of intense solar activity, would alleviate this hazard. He was right, of course, and the radiation experienced by the astronauts on trips to the moon was of no medical significance."

In an effort to expedite a man in space soonest, Low and the rest of the panel spelled out the need to leverage the most reliable booster system available.[38] They recommended taking current, shelf-ready rocket technology such as the Atlas and the Redstone. They also planned to use Faget's ballistic, or blunt-capsule, design. They chose this approach over competing proposals for more elegant winged designs, such as those in use with the X-Plane program, which would have been more in-line with the von Braun paradigm. While this later winged design would influence Low and NASA in early proposals for the future space shuttle in the late 1970s, it presented too many practical, developmental, and budgetary challenges in the 1960s that precluded its use. With cost and speed being of paramount concern, Low began tweaking the von Braun paradigm in order to ensure that America's first efforts in space could get funded and realized.

Simple and to the point, the final plan as submitted sliced through all the competing arguments and interagency noise, and it laid bare a straightforward project in easy-to-understand terms. "Low could write, which is not common in an engineer," Gilruth said, when praising Low's many contributions in the Office of Manned Spaceflight.[39] "It meant he could think clearly." And given the many competing ideas at play in the panic following Sputnik, clear thinking was a critical necessity.

As explained in *This New Ocean: A History of Project Mercury*, "Although the Air Force, Army, and Navy, as well as numerous aviation industry research teams, also had plans that might have worked equally as well, the nation could afford only one such program. The simplest, quickest, least risky, and most promising plan seemed to be this one."[40]

Low ultimately viewed the current man-in-space-soonest goal as just the first, iterative step in a much larger, long-range vision for space exploration. Consistent with the von Braun paradigm, he was already thinking about going beyond simple Earth-orbit missions. As early as 9 April 1959, in the U.S. Senate before the Committee on Aeronautical and Space Sciences, Low outlined his long-range vision. The committee was chaired by Lyndon B. Johnson, who was then a senator and a big supporter of the space program. Low was speaking in support of the bill to authorize the more than $485 million being requested by NASA for fiscal year 1960 for manned spaceflight efforts:

> To put things in proper perspective, I would first like to outline the steps, or milestones, in our long-range program. I have listed these on a chart. At the top of the chart, we have the X-15 research airplane, which is now being readied for its first flight tests. The first orbital flight will come at the culmination of Project Mercury, which is our manned satellite project.... Although Mercury constitutes a logical and perhaps the only first step in our manned exploration of space, we are already studying more advanced systems. I refer in particular to satellite vehicles that will have the capability of maneuvering once they reenter the atmosphere so they can land at a small preselected site. Then, as more advanced boosters become available, we will develop the capability of sending several men into orbit for longer periods of time than one day. We will then be in a position to conduct scientific experiments in a series of manned orbiting laboratories. The ultimate such laboratory will be a permanent manned space station which will be resupplied periodically from the ground. Concurrently, we will make efforts to fly further away from Earth, perhaps to the vicinity of the moon; but a landing on the moon must await the development of boosters that are nearly twenty times as powerful as those that are available today.[41]

Low's testimony is striking in his efforts to continue to promote the von Braun paradigm, when one considers that at the time, there were no plans for anything beyond the accelerated Project Mercury program. Behind the scenes, he was an early and passionate advocate for human lunar landing missions within the internal NASA planning meetings. They certainly had no presidential mandate, congressional authorization, or funding for such efforts. In fact, Glennan admitted many years later that at the time, he was not all that excited by human spaceflight. He was more in favor of satellites and purely robotic programs. Worse, the current Eisenhower administration viewed Mercury as a terminal program; they were very skeptical of human space exploration's massive budgetary considerations.

"Before Alan Shepard's successful flight, most of us were very deeply involved in Mercury," Low said in describing the time. "We worried about the day-today details of the program, and our future was tied completely to it. We did not know whether the country would support a major effort beyond Mercury." In his congressional testimony, he was operating in a virtual political and policy vacuum—guided by his own personal best judgment on technical and creative engineering.[42] He felt that a stretch goal such as a lunar landing would force the greatest advancement in the nation's technical capability and provide the best return for the money, effort, and risk. While he pushed off the timing of such a program well into the future, he felt it important to place the current program in that kind of a long-range context. Despite the terminal nature of Mercury, a lunar landing mission would remain one of Low's primary, driving goals for NASA.

***

Low and the panel presented their findings five days after NASA was officially born.[43] While the Space Act was passed by Congress earlier in the year on 28 July 1958, NASA didn't become fully operational until 1 October 1958. Glennan was impressed by the team's quick ability to come to a consensus, Low recalled. When asked what Glennan's response was to the briefing, he said it was as simple and straightforward as their own conclusion: "Get the hell to work."

Gilruth went back to Langley and established what was to become the STG, charged with operationally bringing Project Mercury to life. "Bob went to the Langley Research Center and told them the number of people he needed," Low said, though they would get "a few from Lewis also."[44] Initially, Gilruth had no specific assignments to fill. He had simply a proposal and a broad idea and a very enthusiastic team. Like his PARD before it, the STG operated in an environment very much like a modern-day Silicon Valley start-up—staffed with young, bright, ideological overachievers like Low, hungry to make the next big breakthrough. "George was always at the heart of the action," Gilruth said. "I was so impressed with him that I asked him to join me in managing Project Mercury."[45]

"I spent about two weeks at Langley working for Bob," Low recalled.[46] Gilruth had offered him the tantalizing job of being Faget's deputy, working on spacecraft design and hands-on engineering. It would have been a dream job for Low. But it wasn't meant to be. "Our boss needed him more in Washington," Gilruth explained.[47] The field, where Low longed to be, would just have to wait. A disappointed Gilruth notified Low one afternoon that he'd gotten a rather surprising call from Abe. "He wondered what happened to you," he said. Surprised, Low told Gilruth not to worry. He would go to NASA Headquarters and remind Silverstein of his promise to give him a field assignment. But Silverstein, very loosely paraphrasing the historic command of Alexander Graham Bell to Thomas Watson, told him: "Yes, but I need you, George. Come for a short period of time." Loyalty overshadowing passion, Low agreed. As this "short period of time" began to stretch out into months, it was looking as though it would extend out into years. Low and Mary R. eventually packed up the family from Cleveland and built a house in Bethesda, about a half-hour commute from downtown DC.

"I've always done what people wanted me to do in this business," Low said about his career and career choices and his ultimate decision to forego fieldwork for an administrative post in Washington.[48] He believed a career was not always a linear plan; it was best to have the courage to seize unique opportunities when they presented themselves. Helping to give birth to NASA and set its long-range plans and goals was about as unique an opportunity as he could find at the time. It was an offer he could not refuse. "I could have said no to any number of things," he explained.[49] "But the forks in the road led a certain way, from the first one that led me to stop in Cleveland instead of going out to Ames. That was a toss of the coin almost. If I had gone out to Ames, I never would have met Abe, and he never would have asked me to come to Washington." He also was not title conscious. The nature and importance of the work justified the job. "I never thought beyond the immediate job, like where am I going to be and what I am going to wind up at the end of it. I've always been driven by the challenge of the job that I have been offered at that time."

Low's philosophy also reflected his unabashed optimism that the road less traveled could also realize unique and brilliant outcomes, none of which could be predicted at the start. It is perhaps a thought planted in his mind as a young boy, traveling to an unknown future aboard a large vessel guided only by faith in the future and following along the path of a brilliantly lit star field. It's career advice he often gave not only to his own children but also to countless students at RPI and colleagues at NASA over the years. He liked to quote the following lines from the noted American scientist and essayist Lewis Thomas, with a particular emphasis on the "what the hell, let's give it a try" sentiment:

We are not like the social insects. They have only the one way of doing things and they will do it forever, coded for that way. We are coded differently, not just for binary choices, *go* or *no-go*. We can go four ways at once, depending on how the air feels: *go*, *no-go*, but also *maybe*, plus *what the hell, let's give it a try*. We are in for one surprise after another if we keep at it and keep alive. We can build structures for human society never seen before, thoughts never thought before, music never heard before.[50]

Making the go decision, he threw himself into his work. He learned all he could about the people and the projects under his charge. His hardworking nature did not go unnoticed by his colleagues or the press. "There was nothing halfway about George Low," George W. S. Abbey said of his work habits.[51] Abbey worked as Low's technical assistant after the tragic *Apollo 1* fire. They forged a close working relationship and a deep friendship. "He was at work long before most people came in the morning and long after they left at night." A reporter for the *Washington Post* once noted during Project Mercury that Low would be in his office at five in the morning and would not leave until nine or ten every night: "Low reads every piece of paper that goes through his office. He feels he has to touch everything to assimilate it. But once he does, he never forgets it."[52]

Low was instrumental in shifting the old NACA academic culture of slow, methodical research at NASA into one of fast, operational efficiency. His penchant for punctuality and not wasting time became legendary. "He was famous for being on time," said Jerry Bostick, a NASA engineer and future deputy manager of the space shuttle program, who served earlier in his career as an assistant in NASA Headquarters.[53] Low's demand for punctuality was a tough adjustment for some, and he had a simple, effective trick to get everyone in-line. "If a meeting was scheduled for ten o'clock, it damn well started right at ten o'clock," said Bostick. "That's when George would lock the door."

No matter how hard someone pounded on the door, they would not get in. Few were ever late a second time. Max Faget found that out the hard way one time, according to Bostick. After being ten minutes late for an important meeting, no matter how hard he pounded on the door, Low would not let him in. He was on time in the future. There was also no reason a general-update meeting should go an hour. He told Bostick to schedule all his regular meetings for forty-five minutes and to block out the last fifteen for phone calls and to take his own personal notes. If a meeting was done in thirty minutes, that was a bonus. Conversely, if a meeting was going long, the opposite was true. Original STG member and Langley engineer Chuck Mathews found this out the hard way. According to Bostick, Mathews was making a presentation to Low, and his delivery was going long. After thirty-five minutes, Mathews was still nowhere near his conclusion. Low, watching the clock carefully, cautioned him: "Chuck, if you've got a point to make, you better hurry up and make it in the next ten minutes." Chuck nodded in the affirmative, but he kept "plowing along in the same manner." Fifteen minutes later, Low slapped his notebook closed as Mathews was somewhere in midsentence. Without saying another word, Low simply stood up and left the room. The meeting was rescheduled. "Chuck, now you've experienced it. You know what to expect," Bostick said to a stunned Mathews. "I recommend, next time, you *start* with the conclusion, and go on from there."

Mathews nailed it the second time around.

Given his frenetic and jammed schedule during the week, Sundays became sacred to Low.[54] "I try to make it a rule not to work on Sundays," he said in a profile interview during the Apollo program.[55] "I keep Sunday for going to church and playing with the kids." In media interviews, he often expressed his appreciation to his family for their support and the sacrifices they made on his behalf, and especially to his wife, Mary R. Case in point, when Low was given the Arthur S. Fleming Award as one of the ten most outstanding young men under forty in government in 1963, he told the audience that Mary R. was the program manager for one of the most challenging yet thankless projects in the city.[56] The job? Taking care of Low and their five children. "My frequent trips out of town, and my even more frequent evenings and weekends at the office, have not made her life an easy one," he admitted. "Yet she is always cheerful, friendly, and uncomplaining."

When not traveling, Low would rise early on Sundays and reconnect with life at home. His typical, family-focused Sunday was described by writer William Barry Furlong: "Low has five children . . . and they adore the elaborate manner he does everything for them on Sunday, from water-skiing to serving up pancakes. 'He makes spectacular pancakes,' says his wife, Mary R. In fact, he also makes spectacular scrambled eggs Benedict. 'I won't try to compete with him on making any of those breakfasts,' says Mary R. In cooking, if not in engineering, Low indulges a certain spontaneity. 'He's long since forgotten the recipes.'"⁵⁷

He loved to recharge his batteries with the family in the sun—whether water skiing, jogging, bicycle riding, or playing tennis.⁵⁸ As in his NACA days, he enjoyed the challenge of working on the family car or restoring old hobby cars—the kids serving often as his assistant. Despite the focus on family, work still was never very far off. It would often manifest itself in the form of telephone calls or stacks of memos to read, decisions to make, and notes to take and disseminate. Low carried with him two briefcases at all times, one with incoming memos to read and the other with read and notated memos and outgoing correspondence. If any given weekend afternoon was prone to good weather, he would go out into the backyard in a lawn chair to read, his trusty green felt-tip pen in hand. A long extension cord would snake out from the house to a record player for background music. His two briefcases were opened on either side of him as he made quick, methodical work of catching up on the never-ending onslaught of NASA agency memoranda.

In Washington, Silverstein appointed Low head of the Office of Manned Spaceflight, making him one of the agency's first official employees. "I was there the first day there was a manned space program," Low pointed out.⁵⁹ His team was small, but influential. In addition to administrative support, Low brought with him from Cleveland John Disher, as chief of advanced manned spaceflight, to focus on future space applications, and Warren North, as chief of manned satellites, who would oversee STG operations in Langley and then in Houston for Project Mercury.⁶⁰ A few years later, when overseeing the planning and start-up of both Gemini and Apollo, his team would only expand to ten handpicked people. In contrast to Apollo, where he managed a staff of over fourteen thousand people in Houston, the early days of NASA Headquarters were lean. Staff buildups focused out in the field centers where they were needed. "When NASA was established, George Low and Warren North *were* manned spaceflight," John Disher said.⁶¹ "I was working on the Space Task Group on Mercury out of Cleveland, and about to move to Langley, when George called in the spring of 1959 and brought me to Washington. With Eisenhower and Glennan not committing to any program beyond Mercury, I *was* advanced manned missions for two years." Together, they ran the show.

In addition to his work in Washington, Low would spend a considerable amount of time airborne, traveling among NASA's various installations for meetings and progress updates.[62] He would go down to Cape Canaveral in Florida and back to Wallops in Virginia for test launches and more meetings. He became the hub of all the spokes, and he took his role seriously to make sure he knew every important detail, from personnel through to the performance of the spacecraft. Of course, he relished the opportunity to dig into the engineering whenever he could. "George and I would spend a good two or three hours every evening, going over every detail of what had gone on during the day," recalled Rodney G. Rose, a project engineer on the Little Joe rocket at Langley, which was being used to test the Mercury spacecraft in the early days.[63] "George was a meticulous engineer, and we reviewed every detail on the capsule, the booster, and everything else."

Additionally, as head of human spaceflight, Low would be on the front line dealing with Congress and the media. In no short order, he would find himself giving public speeches, answering questions from a voracious and ever-prying press, testifying before Congress, and helping to not only guide the program through its rocky start but to also begin to plant the seeds for the next phase of NASA's long-range plan. "There was a time when I spent 90 days testifying before various committees of Congress," Low said.[64] Throughout his professional career, a reporter noted, he would always shine a bright, focused light on the space program in press interviews and in congressional testimony.[65] He never focused that light on himself. The work was never about him. It was always about the details—in the work, on the team, and with the goal. "He's not the kind of guy who pushes himself into the limelight, but without him, the program wouldn't be," a headquarters colleague explained in the same article.

When the Russians continued to beat NASA to a series of manned spaceflight firsts during Mercury, it was often Low who stood in front of Congress dealing with the outraged aftermath. He was quoted in the media, reminding the nation that America's rapid progress and long-range plans were the right ones to follow. He cautioned against short-sighted and defeatist thinking. "Flight into space is just too exciting an adventure to miss," he liked to tell people.[66]

***

The young engineers working for Gilruth were all classically trained aeronautical engineers and not yet experts with missiles and rockets. Despite the simplicity of the report by the Manned Satellite Panel, they had a lot of ground to cover and applied research to perform in order to catch up to the progress of the Russians. To facilitate the work and NASA's progress, Low operated as the STG's man in Washington and as a go-between for Silverstein and Gilruth. He traveled back and forth between Langley and Washington and from Washington to the various NASA installations across the country. He coordinated the growing and varied theoretical research projects and application tests, such as Knauf's space medicine research projects, and documented the program's progress via regular status updates for Gilruth, Silverstein, Dryden, and Glennan. He would make a career out of being the strong number two man—the ghost in the machine who got things done.

A considerable portion of his job became one of administration and operations oversight. According to Low, the STG was a highly technical organization filled with expert engineers who showed little interest in the business management aspects of the program.[67] "Because of this lack of attention to the business management function, particularly financial management, we had to perform this task out of my office. STG just didn't have the capability to handle it in the early days," Low said. Nonetheless, his role in this regard would not be insignificant. He testified before Congress in support of important budgetary appropriations, as well as helped develop the managerial and oversight infrastructure that would form the core of the human spaceflight effort.

No job was ever too big or too small for him.

The one area of personnel management that was handled purely at the STG level and not in Low's office was that of astronaut flight assignments. This was handled in the early days by Bob Gilruth. Although Low was a central team member in the selection of the Mercury Seven astronauts, Gilruth had the final say in who would launch and when.[68] He would sometimes elicit input from Low or from his two deputies, Charlie Donlan and Walt Williams. But according to Low, Gilruth mostly relied on peer ratings from the astronauts themselves, coupled with his own personal judgment. "Bob asked each of the seven astronauts, 'If you were unable to fly, assuming you wanted to fly first, but were unable to, who would you like to see fly next? And why?'" Low explained.

There were, of course, difficult and rare exceptions to this peer review process, such as the medical groundings of both Deke Slayton and Al Shepard. "I was one of two people who told Al Shepard that he would never fly again after his first Mercury flight," Low said.[69] He and Gilruth had met with Dr. Charles A. Berry, who was part of the original astronaut selection committee and the medical monitoring physician for the Mercury program. Berry told them that Shepard had Ménière's disease, an inner ear disorder that induces tinnitus, fluctuating hearing loss, and vertigo. He also told them that it was incurable. Shepard was devastated. "I remember Bob had a yellow couch and Al was sitting in it and we were sitting on the two sides," Low recalled. "It was one of the most difficult things I have ever participated in. Telling a guy that what he's wanted all his life, he won't be able to do, is very, very difficult." It was equally as hard when they had to ground Deke Slayton for a heart condition. Of course, both men would later go on to fly successful space missions, and it gave Low great personal satisfaction to witness his two colleagues overcome seemingly insurmountable obstacles and achieve their dreams.

Low, a licensed pilot, had a special affinity for the role of the astronauts and the technical quality of the individual who should be entrusted with either flying or commanding a mission as risky as spaceflight. He was influential in lobbying for early decisions in Washington that would keep the focus on selecting highly skilled and experienced engineering test pilots. He was part of the team that selected and recruited the experimental test pilots who would become the nation's first astronauts. "The military was principally involved in selecting the first pool of potential astronaut candidates," he said.[70] "They picked the first group of 508, then cut it down to 110, and to 70. Then NASA took over at that point."

In 1959, small groups of test pilots were invited to Washington for a briefing on the project and to be asked to volunteer. Low and Silverstein would brief them.[71] They had a tough crowd to pitch to. There were, of course, the derisive references by some of the pilots—such as Chuck Yeager or Scott Crossfield and others at Edwards—about the Mercury astronauts being nothing more than "spam in a can" because of the highly automated nature of the spacecraft.[72] In their minds, the astronauts were not true pilots in full control of their craft. This derogatory notion was not helped by McDonnell Douglas engineers using Yorkshire pigs as live test subjects during high-altitude drop tests of the original Mercury capsule. Low didn't appreciate or agree with the reference. After all, the cosmonauts who would go on to fly the celebrated Russian Vostok would not have the same kind of piloted control over their craft that the Mercury astronauts would. From Low's perspective, if anyone was pure spam in the can, it was the Russian cosmonaut, not the American astronaut. Regardless of degree of magnitude between automation and manual control, Low was convinced that test-piloting skill and experience during Mercury would be crucial in the outcome of a successful program. From a recruiting perspective, he believed that emphasizing this point was potential catnip for attracting qualified volunteers.

In *The Right Stuff*, Tom Wolfe chronicles the nature of Low's briefings as he underscored the astronauts' role in a mission:

> The slender Low went out of his way to show that the astronaut would exercise some forms of control. He would have "flight attitude control," for example. In fact, this meant only that the astronaut could make the capsule yaw, pitch, or roll by means of little hydrogen-peroxide thrusters, just as you could rock a seat on a Ferris wheel but couldn't change its orbit or direction in the slightest. But when a capsule was put into earth orbit, said Low, controlling the attitude would be essential for bringing the capsule back in through the atmosphere. Otherwise, the vehicle would burn up and the astronaut with it.

Low also described the fly-by-wire system, in which the astronaut could override the automatic control system and take manual control. This was especially important in the case of an in-flight anomaly or malfunction. It was a heavy sell by Low, but he meant it.[73] He knew that experimental test pilots were offered what Wolfe characterized as "innovative duty," or "an ill-advised tangent." These were missions that could not only take them out of favor within the hierarchy of military command at their respective flight centers but also put them further behind in jet flight-test time and pole position for military promotion. Fear of innovative duty could be a deal breaker. As such, Low assumed that it would be an uphill battle. After all, not everyone subscribed to the George Low carpe diem style of career management. The pilot response, however, astonished him. Wolfe explains, "From the beginning, George Low and others in the NASA hierarchy . . . were amazed. They had briefed thirty-five test pilots on Monday . . . another thirty-four the following Monday; and of the total of sixty-nine, fifty-six volunteered to become astronauts. They now had so many volunteers they didn't even call in the remaining forty-one men who fit the profile. Why bother? They already had fifty-six grossly overqualified volunteers."[74]

Low remained adamant that the success of any given mission depended on the skilled action of the pilot, either in performance of a primary function or in the performance of a backup function. As incoming experienced test pilots, the astronauts' learned situational awareness and practiced ability to react in the most adverse, unanticipated situations would be honed to a high level of expertise. "Although the Mercury vehicle could fly completely automatically, the pilot will play a very important role in the mission," he told the House Committee on Science and Astronautics on 4 May 1959.[75] "Mercury is essentially a flying machine. If anything should go wrong, the pilot has the best capability of correcting this."

Walking the House committee through the pilot's role, he pointed out that it was the astronaut who would communicate with the ground stations, make scientific observations, and monitor all the onboard equipment and make active corrections as required. More importantly, he said, the pilot could control the attitude of the capsule, changing its position, and also navigate against the stars with a built-in periscope. It was no mere sales pitch. The astronauts, he said, were truly piloting the craft.

Low testified, "Since he can control his attitude and he can navigate, he can also fire the retrorockets. He can do this even if he should lose all contact with the ground. He can initiate other emergency procedures, such as activating the escape system, and he can deploy the landing parachute if it is not deployed automatically. Because these functions are so very much like the functions that engineering test pilots perform in their everyday life, we limited participation in the program to this group."[76]

As a matter of fundamental engineering philosophy, he also believed that any user of a mechanical system should also be intimately involved in its design. The astronauts, therefore, would be elemental to the overall engineering design of the capsule and its control systems, too. In testimony before an executive session to the same House committee two weeks later, Mercury astronaut Wally Schirra confirmed the astronauts' role as Low had planned. Schirra described how the astronauts provided input and suggestions for critical engineering changes to the overall spacecraft, which was being manufactured by McDonnell Douglas in St. Louis. After Schirra spoke, Low told the committee, "This suggestion of changes by this group of astronauts was not accidental. We had planned in our program on taking the astronauts to McDonnell to look over the capsule and to tell us specifically how the pilot display and how the cockpit layout should be handled because we feel they are experts in this area. They are therefore not really making changes. They are responsible for the design."[77]

"The astronauts were a part of a team," Low clarified.[78] "The astronauts were deeply involved in the design, in the mission planning, from the very beginning. I think one of the reasons for the success of Mercury, Gemini, and Apollo is that the astronauts were involved in designing the flight systems. They understood them; they knew that they could manage them. They developed the flight procedures, then had the discipline to abide and stick to those procedures. They were just a totally integrated part of the team, and that, I think, was a very important aspect of the whole program. They were the hardest-working bunch of guys I ever knew."

Not everyone agreed with NASA's astronaut criteria, especially in Congress, where other political and societal considerations often took precedence. The role of the astronaut had taken on celebrity status, and a myriad of people wanted a shot at being one. A member of the House committee, Representative James Fulton of Pennsylvania had a habit of getting under everyone's skin. He often picked fights on technical and side matters to the point of absurd annoyance. For example, Fulton would continue to press Low in open session on the perceived fairness of the selection process of the original Mercury Seven astronauts and the qualifications necessary for selection, despite the fact that they already had more than enough astronauts for the few flights they anticipated and despite the fact that the president and Congress had not mandated any further human spaceflight program beyond Mercury. A veteran of the navy, Fulton questioned the lack of army representation in the ultimate selection.[79] After all, the army had its own advanced space planning and rocket initiatives. Of the original Mercury Seven, Fulton noted, three were from the navy, three were from the air force, and one was from the marines. None were from the army.

Low, himself an army veteran, would remind Fulton and the committee that although the army had a rocket program, it did not have a test pilot program.[80] They were field soldiers. They did not fly jets. However, the army was involved in the medical evaluation of the target pool of volunteers, and the army was of course providing critical support in rocket development in Huntsville. Although not represented in the group of astronauts, the army was still a major part of the process. "In most of the selections we did not work with the names or services," Low pointed out. "We assigned numbers . . . and didn't really know until after we had selected the final seven what service they were in."[81] Fulton, not satisfied, would continue to press the matter during testimony. Low did not let the politics of the chamber or its proceedings rattle him. He simply stayed on message and communicated in his calm, well-reasoned manner:

> MR. FULTON: As a Navy man, may I say you certainly have the standards rigged so you effectively dealt the Army out for a big zero on the honor of getting first into space. Why is it necessary to be a qualified jet pilot other than to pull a couple of levers when emergencies might arise? Nobody has ever been in space before.

> MR. LOW: No, but there will be many unusual situations we can't forecast. This machine is flying at 18,000 miles an hour through space. It may become disoriented through, for example, the last kick of the rocket. The experienced jet pilot gets into situations every day where he is not oriented and he has the knack, and the qualifications, to immediately right himself again. . . . He has to operate a control stick which is very similar to the stick on the jet airplane. We felt that in order to insure the success of the mission, we should consider only the qualified engineering test pilot, the jet pilot, for the first mission. Later on, as we learn more about spaceflight, I am sure we will be able to relax our standards. But I should also say that we did not intend to favor one of the services over another one. Our main purpose was to make the mission a success.[82]

The program's other main purpose was to achieve the goal of manned spaceflight in the shortest time practical. At the time, this also excluded opening the selection process to a broader scope of American society. This decision unfortunately excluded women and minorities or any other pilots who were not part of the U.S. military's advanced test pilot program at that time. Low made it clear to the committee that gender and racial bias were not built into the selection process.[83] Their intent was not to be overtly exclusive of any socio-economic or demographic groups, just expedient in the skill sets required. They were operating under the need for specific capabilities and skills necessary to achieve the goal as soon as possible. Relying solely on volunteers from the existing experimental test pilot schools of the military, they came up with a list of 110 qualified applicants for less than a dozen needed slots.[84] With the limited number of flights in the program and no follow-on program or opportunities as yet planned, they had more than enough pilots in the pipeline. In fact, Low's initial recommendation was to select twelve astronauts; Dryden, however, pared the number down to seven, mostly because of the limited number of flights.[85]

The nation's first official astronauts, the original Mercury Seven, were announced on the afternoon of 9 April 1959: Virgil "Gus" Grissom, Leroy Gordon Cooper Jr., and Donald "Deke" Slayton, from the U.S. Air Force; Alan B. Shepard Jr., M. Scott Carpenter, and Walter M. Schirra Jr., from the U.S. Navy; and John H. Glenn Jr. from the U.S. Marines. House Representative B. F. Sisk of California spoke up on behalf of NASA to support the approach Low and his team took in selecting the seven:

> Of course, I have a great regard for these gentlemen, and realizing the tremendous responsibilities, I am inclined to disagree with my good friend from Pennsylvania [Representative Fulton]. I think these gentlemen were confronted with a practical problem which make it utterly impossible to go through a long series of tests to test every individual in the United States who might have sought this examination. We have a practical problem. We are demanding urgency, expeditious action on the part of these people. I think they have done a wonderful job here and I want to commend them on it.[86]

Still, the issue of selection would again reassert itself. On 8 February 1960, while going through a technical update on the state of the program, Fulton interrupted Low. In the fall of 1959, *Look* magazine hired Betty Skelton, a thirty-three-year-old aviatrix, as part of an admitted publicity stunt to participate in some astronaut exercises at a U.S. Air Force base in San Antonio, Texas.[87] The editors of *Look* titled the cover story "Should a Girl Be First in Space?" The article caught Fulton's attention.[88]

MR. FULTON: One other thing. There are always rumors getting started. I read an article that a woman may be first in space. Is that wrong? Are you training or recruiting any women for this?

MR. LOW: I can only speak for this nation's effort. [Chamber laughter.] We have only the seven astronauts who were introduced to you gentlemen last spring, I believe. We have no additional ones.

MR. FULTON: I saw a very comely young lady on the front page of one of the large circulation magazines in the United States. It stated that she was being considered.

MR. MCDONOUGH (representative from California): That was for a newspaper magazine story.

MR. FULTON: Is it true or not?

MR. LOW: She is not being considered by us.

MR. OVERTON BROOKS (committee chairman and representative from Louisiana): The rumor I heard was that a congressman was going up.

MR. MOELLER (representative from Ohio): I heard he was going to be the chairman of our committee.

MR. FULTON: I volunteered two years ago and was rejected. [Chamber laughter.]

A few years later, when newspaper stories started to appear out of New Mexico revealing that a private clinic was testing women pilots to become astronauts, Fulton called a special committee hearing to investigate. Dr. W. Randolph Lovelace, who conducted the medical tests on the original Mercury Seven astronauts, decided on his own in 1960 and 1961 to run a series of experiments to see if women pilots could meet the medical standard to qualify.[89] Low, John Glenn, and Scott Carpenter represented NASA in the hearings. Several women pilots from the Lovelace Clinic experiments were also asked to testify. In his testimony, Low defended NASA's position, explaining (again) that necessary astronaut requirements at the time unfortunately meant that no women yet qualified. The additional time and expense of testing, selecting, and training not only women but other pilots outside military programs and their qualifications already considered would unduly slow down the effort in their race against the Soviets. This was something Congress did not want NASA to do. Low argued that solving the nation's many social issues was not in NASA's mandate; beating the Russians into space, however, was.

Low, of course, was not an unsympathetic victim of the same rules that kept the other pilots—women or men—from being selected as an astronaut. Although he did not express it in his testimony, Low harbored a professional desire to fly into space.[90] Not getting that chance was a professional regret upon retiring from NASA in 1976. "There is nothing I would have liked more," Low said in a post-NASA interview.[91] "The hardest thing is being able to do everything up to that point, then not being able to make that flight."

As Low stated many times, he felt that in the future, if America's space-faring capabilities advanced and were adequately funded—especially with a program like the shuttle—the door would swing wide open for a much more diverse and inclusive group of astronauts. This wasn't just idle political spin. He meant it. It is a sentiment he shared later in his career with some of his very close NASA colleagues—some of whom actually had the privilege of flying into space. Upon Deke Slayton's successful participation in the Apollo-Soyuz flight in 1975, when the space shuttle was very much on the drawing board and moving toward reality, he presented Low with a flown, silver medallion as a memento. Along with the medallion, Slayton wrote Low a personal note touching on his knowledge of Low's desire to expand access to spaceflight:

> The enclosed ASTP [Apollo-Soyuz Test Project] medallion is a small token of my appreciation for all your efforts on my behalf over the years. In particular, there is no way I can thank you adequately for your part in assigning me to the Apollo Soyuz crew. It is an experience in which few have been privileged to participate, and I doubt that anyone has ever enjoyed it more. Our many associates who have been ground-bound for years deserve an equal opportunity. I see the Space Shuttle as the vehicle for providing many of these folks, but most important, yourself, that opportunity. I will work diligently towards the goal of making it possible for you to personally view the great beauty of Earth from space.[92]

Low also made this argument when, as NASA deputy administrator years later, he worked to open the ranks of both the astronaut corps and engineering staff to more women and minorities. Despite his comments and his personal beliefs, it would be the stunning and dramatic testimony of a celebrated female pilot that would eventually take the air out of the room and

Fulton's argument. The woman was Jackie Cochran, a fifty-six-year-old pilot who personally paid for most of the women to take the tests at the Lovelace Clinic.[93] In *Integrating Women into the Astronaut Corps: Politics and Logistics at NASA, 1972–2004*, author Amy E. Foster points out that Cochran was a bit of a tainted witness in that she "felt herself the most qualified to be the first woman in space," but her "advancing age and declining health made her selection impossible."

Nonetheless, Foster writes that Cochran blindsided her fellow women pilots with her testimony when she agreed with Low: "I do not believe there has been any intentional or actual discrimination against women in the astronaut program to date." The inclusion of women in the space program, Cochran said, "should not depend on the question of sex, but on whether it will speed up, slow down, make more expensive, or complicate the schedule of exploratory space flights." While the hearings and the formal debate in relation to Project Mercury were brought to a close, the issue itself would continue to haunt NASA management. It would come to a boiling point in the mid-1970s, amid the broader equal-opportunity mandates legislated by Congress. For the time being, however, it was still a Man-in-Space-Soonest program. Indeed, only a Seven-Men-in-Space-Soonest program.

***

During the course of Mercury, Low was concerned that the public and the media would view America's space efforts as purely a race with the Russians. Knowing where the United States was on the development curve and in context of the agency's overall, long-range ambitions, Low cautioned Congress and the press against using the emotionally effective yet potentially limiting metaphor. Despite the current push by some in Congress to beat the Russians into space, he said the first manned Mercury launches were target dates only. A final launch decision would depend on the progress of further testing. And given America's slow start, there was a very good chance the United States would not be first.

"There are many new developments that we must complete before Mercury can be brought to its fulfillment," he adamantly told the committee as early as 1959.[94] "Although we are trying to stay within the present state of the art as far as possible, we still must perform many tests on the various components.

We may, at any time in the program, run into a difficulty that is now unforeseen which may delay us by a week, a month, or perhaps even several months. But currently, we are planning to use the solid rocket cluster, the Little Joe vehicle, for our development tests during the last half of calendar year 1959." After the flight-ready McDonnell capsules became available, he explained, the astronauts would first gain actual flight experience doing a series of ballistic shots with the Redstone booster, which would give them short, four- and five-minute periods of weightlessness. "To qualify the pilot and to train him, we believe he should undergo some of the stresses he will be subjected to in an orbital mission. These include launching accelerations, the decelerations during reentry, the use of the attitude control and the stabilization systems, and a short period of weightlessness. All of this can be achieved with the Redstone boosted capsule."

After these test flights, then, and only then, he said, would NASA attempt the first orbital missions using the modified Atlas ICBM booster. Despite Low's objections, animal flights would also be a part of the testing regime.[95] Dogs, pigs, and monkeys, after all, are not used to test new airframes on their maiden flights; why should they be used with spacecraft, if the planning and testing regimes are done carefully enough? At the time, it was felt that there were too many unknowns and concerns, especially from the medical experts, about whether or not a human could survive a trip into space.[96] Even von Braun and many of those developing NASA's rockets in Huntsville agreed with the animal-testing approach. Low went along with the program. "After instrumented flights are successful," Low said, "we will fly animals in satellite orbit with the Atlas." Finally, after all those tests had completely proven the reliability of the system—as well as the ability of a living creature to survive in space—then the astronauts would be allowed to fly an orbital mission.

Despite Low's cautions, House Representative Leonard G. Wolf of Iowa suggested that the agency put its budget request and program into terms of psychological warfare in the battle against the Russians to be first into space. "I think we need a strong statement here, because we are talking now about $70 million," Wolf said, expressing a perennial NASA debate over cost and benefit, as well as public perception.[97] "There are a good many people in this country who still think this is all foolishness. But they will understand psychological warfare."

Low reminded Wolf and the committee that simply getting a man into space should not be the only goal for NASA. "I think we should recognize, however, that we may not be first," he repeated.[98] "But even if we were not first, this is an important step and perhaps only the first step in our long-range manned spaceflight program. This is going to be a continuing program that leads to space stations, to interplanetary flight stations."

Olin "Tiger" Teague, a representative from Texas and an important, influential political figure in NASA's future, agreed. "Possibly, we are a lot more interested in an individual and his life than other countries," Teague suggested to his colleagues, putting a little political topspin on the point.[99] "We will be a lot more convinced of his return before we send him up, than some other country."

Teague, a hefty, much-decorated World War II veteran, would spend his career on the House Science and Astronautics Committee, befriending the astronauts and placing the text of letters, news articles, and speeches favorable to the space program into the *Congressional Record*.[100] Low would work closely with Teague over the years to advance America's space program, especially during his tenure as deputy administrator of NASA. While they wouldn't always see eye to eye, Teague would become one of Low's go-to congressional allies in the future.

To bridge Teague's and Wolf's positions and to conclude testimony, NASA submitted the following statement on the urgency of Project Mercury, the final language of which Low, as chief of manned spaceflight, would have been responsible for crafting or, at the least, would have used his green hornet to heavily edit:

> One of the most significant milestones in the history of space flight will be the first successful launching, flight and recovery of a man in space. The event will be of such importance that it cannot fail to earn worldwide acclaim for the technical skill of the nation first accomplishing the difficult task. Project Mercury, the current manned space flight program of the NASA, represents this Nation's effort to place man in space. The project was conceived and is being carried out in a manner that will attempt to achieve the earliest possible space flight by man. It will use developed booster systems through all phases of the program. The space capsule, the actual flight vehicle, is based upon design prin-

ciples that require no major technological breakthroughs for successful accomplishment. Even though we in the United States apply our greatest efforts to this problem, we must recognize that we may lose the psychological battle to be first. We cannot place prestige above life. A fundamental concept of Project Mercury is that the safe return of the astronaut must be insured. Other nations having less regard for human life may shortcut the development procedures that we think essential and may thereby accomplish manned space flight before we do. Although we in the United States have no certain knowledge of the desires and intentions of other countries to be the first to achieve manned space flight, we cannot relax our own efforts in the hope that others will do likewise. To do so and to be wrong would insure a certain psychological defeat for the United States in the eyes of the world.[101]

Still, Congress, the press corps, and the public continued to frame Mercury in the feverish context of a sprint with the Russians to be the first to launch a human into space. It was a race the United States would eventually lose. In a buildup to Russia's first human spaceflight, reports out of Washington and NASA began hinting that the Soviets would likely be first. In fact, Low all but assured as such in his press interviews. "I am firmly convinced that the Soviets have the capability of achieving manned space flight," he told the United Press International in the fall of 1960.[102] "I believe they are awaiting only an appropriate time to send their man, or men, into space."

A few months later, in February 1961, nationally syndicated columnist Bob Considine reported that if Soviet technology is *not* the first to send a manned space vehicle into orbit, it will come as a tremendous surprise to nearly everyone working on Mercury.[103] Considine directly quoted Low, the chief spokesman of this understanding, who could not make the point any clearer, even if it wasn't sinking in with the public at large:

> If the only justification for this project was to establish an important first, then it is quite clear that we never should have initiated the program, for it is well known that the Soviets took the first step in a manned space flight program long before 1957. In November 1957, they orbited the dog, Laika, an obvious indication of the general direction of their space effort. Certainly, it is clear that if they were able to achieve Sputnik II in November of 1957, and have an animal alive in the satellite for

a number of days, they must have started this program two, or perhaps more years before 1957. Our National Aeronautics and Space Administration was not even established until October 1958. This was nearly a year after it was evident that the Russians were well underway in a manned space flight program and perhaps three or four years after they had initiated their program. Project Mercury was then initiated by us with a clear understanding that the Russians would most likely achieve manned orbital flight before we would.[104]

While the public and members of Congress took it as a psychological blow when it was announced that Yuri Gagarin had flown first on 12 April 1961, Low and a select group of people within NASA obviously knew better. The nation was no longer lost in a series of uncoordinated programs spread thin among a multitude of agencies. NASA was making increasingly steady progress. They also were gaining rapidly in expertise, experience, and technological capability. Low's statements were an attempt to soften the inevitable blow they all knew was coming. Still, Gagarin's flight was a bitter pill for Low to swallow. Despite his public pronouncements, he was deeply disappointed. He was less consumed by geopolitics and more just a natural born competitor. This competitive streak is obvious in Low's personality, as observed by writer Denise Grady in a later post-NASA profile of Low in *Discover: The Newsmagazine of Science*:

> Low's calm exterior covers an ever-present tension. Heading for an appointment, he walks briskly and determinedly, with a certain stiffness of the backbone. He turns the ignition key in his car before shutting the door or buckling his seat-belt; not a moment wasted. If a gas station attendant does not appear promptly, he is likely to drive off. On the job, if people do not produce after being given time to shape up, he fires them. Don't be late, his colleagues advise, for an appointment or with a report. Don't make the same mistake twice. Don't be afraid to disagree, but don't state an opinion you can't back up. An ill-conceived memo will come back abuzz with "green hornets," a NASA term for the stinging questions penned in Low's infamous green ink. Low is, to the marrow, a competitor. He likes being the best. . . . As an engineer, he is irritated at seeing the United States produce second-rate cameras, television sets, wrist watches, and calculators.[105]

"One of the deepest disappointments to me was when I got a phone call, the night of April 12, at 2:00 a.m., telling me that Gagarin was up," Low recalled, his competitive nature flaring.[106] It was a night in between House testimonies in which Low was updating Congress on Project Mercury. He felt strongly that America had just been cheated out of being first. Shortly before Gagarin's flight, Low's office was ready to launch a manned flight before the Russians. But surprisingly, von Braun and his team in Huntsville insisted on an additional, unplanned animal flight. There had been some earlier test-flight anomalies that had been corrected but not yet flight tested. "I was as mad as George," said Chris Kraft, recalling that bitter moment for those who were ready to launch.[107] The issue at play was a relatively minor one; the risks, minimal. "It was an internal system timing issue for the landing bags, really just a matter of programming seconds, that we had figured out and had fixed. We were ready to go."

There was the political question, of course. NASA was on a short leash. Any loss of life could easily see the entire human spaceflight program canceled. In fact, Low's own push for a long-range plan beyond Mercury caused unanticipated friction within the White House and in political circles that heightened the concern.[108] Members of Eisenhower's science advisory team were piqued by Low's post-Mercury concepts of a manned lunar landing, which Low made in his April 1959 congressional testimony in support of NASA budgetary appropriations, as well as in public presentations he was making for preliminary lunar landing studies. In a knee-jerk response, they formed an investigative committee to review all manned spaceflight activities. Space historian John Logsdon describes this ad hoc committee's formation in *Exploring the Unknown*, volume 7, *Human Spaceflight: Projects Mercury, Gemini, and Apollo*:

> That NASA was planning advanced human spaceflight missions, including one to land people on the Moon, soon came to the attention of President Eisenhower and his advisors as NASA submitted a budget request that included funds for industry studies of the Apollo spacecraft. This request was not approved, and the president asked his science advisor, Harvard chemist George Kistiakowsky, to organize a study of NASA's plans by the President's Science Advisory Committee. To carry out such a study, Kistiakowsky established an "Ad Hoc Committee on Man-in-Space" chaired by Brown University Professor Donald Hornig. The

Hornig Committee issued its report on 16 December 1960. The report called Project Mercury a "somewhat marginal effort," and noted "among the reasons for attempting the manned exploration of space are emotional compulsions and national aspirations."[109]

The impact of the Hornig report was a disaster. The committee informed Eisenhower that landing humans on the moon would cost in the billions.[110] He was incredulous. It all seemed to him to be a silly waste of money. By all accounts, the president and his advisors were bewildered that anyone would even consider such an outlandish undertaking. "This won't satisfy everybody," one advisor said.[111] "When they finish, they'll want to go to the planets." At this comment, it is reported that everyone in the room—including Eisenhower—burst out laughing. Eisenhower's science advisors suggested that the scientific potential of such a program would not be worth the cost. Shortly before John F. Kennedy was to be sworn into office, Eisenhower sent to Congress in January 1961 his budget proposal for 1962.[112] In a parting dig, he questioned the logic of extending manned spaceflight beyond Mercury, and he removed any post-Mercury funding. America's fledgling manned space program looked destined to die on the vine, starved for funding and political leadership—a perennial problem that would continue to haunt and hamper NASA for decades to come.

Despite Eisenhower's concerns on cost, the Hornig committee recommended even more animal flights in Mercury before the first manned launch.[113] The irony of the request was not lost on Low. "They recommended that we could not possibly take the risk of flying a man for five minutes in a weightless flight without many additional tests with monkeys," he said, incredulous at what those additional test flights might cost. "In fact, they wanted us to fly fifty monkeys. We learned there weren't this many trained monkeys available in the country." The animal tests were a complete waste of time in Low's mind. As early as May 1959, he conducted extensive interviews concerning the first American space voyager being completely on his own and ready to accept the risk of the flight. That's what test pilots do, what they sign up for. With "no automatic control from the ground which can take over operation of his satellite," he pointed out, the astronaut "will hurtle spaceward without having first tested the emergency escape procedures" that were being built into the spacecraft. He reminded the reporter that this is in line with procedures currently used in new aircraft, where pilots do not test ejection seats. Their first use of an ejection seat is in an actual emergency. Low also cautioned about booster failure: "The Atlas will not be completely reliable in the time period between now and the time for a first manned satellite launch," he said. "With the addition of an escape system, the flight will be no more dangerous than the first flight of an experimental military aircraft." In his initial briefings with the pilot volunteers, he cautioned the pilots that the flights would be potentially dangerous. That is why they asked for volunteers from test pilots, who were no strangers to dangerous and untested flight duty. There would be risks, they knew; but they would be calculated risks. It's what these guys do. "Gilruth, Silverstein, and I were ready to go ahead," Low said.[114] "We knew what the problem was, and we were sure we had it solved."

But von Braun would not budge. In an attempt to overcome launch objections, Silverstein called for an emergency meeting in Washington. Maj. Gen. Don R. Ostrander, who was director of the Office of Launch Vehicles for NASA in Washington, was responsible in NASA Headquarters for the launch vehicle, and it was ultimately his call. Joachim "Jack" Kuettner, von Braun's director of the Mercury Redstone project, showed up for the meeting. Von Braun, curiously enough, did not. But Kuettner brought along von Braun's concerns and insisted on an additional test.[115] In the meeting, Ostrander sided with the von Braun team. He felt that caution was merited; he especially was concerned with the political fallout should the launch fail.

As Low knew it would, the subsequent animal test flight flew without a hitch. The problem had, indeed, been solved. Before NASA could turn around another capsule and booster for Shepard, Gagarin flew. "Irony: von Braun licked us, not Gagarin," journalist Robert Sherrod wrote in his reporter's notes when hearing this story from Low for the first time.[116] In truth, Low did not blame von Braun or Ostrander. Rather, he blamed incessant political pressure and an improper understanding of the actual risks involved. "It was a political (and not technical) decision," Low clarified. And it was a valuable lesson in Washington decision-making that he would not soon forget.

***

Public opinion can be fickle, and momentum can swing in the blink of an eye. Although Low was personally disappointed at not being first, that disappointment proved to be temporary. He credits administrator Webb and his appearance on the *Today Show* the next day, after Gagarin's flight, "with causing the country not to panic," as well as Webb's masterful handling of the uproar in Congress.[117] "Webb was almost brand new in his job," said Low. He had driven Webb up to the Hill, feeding him information along the way—size of the Soviet payload, thrust of their booster, that kind of information. "I was amazed at how well he absorbed it. Webb assessed what the facts were, and he fed them to Congress just right," he recalled, giving Webb all the credit in the world for pushing back on the Kabuki-theater disappointment coming from the same crowd that was insisting on more animal test flights.

Alan Shepard's historic suborbital flight on 5 May 1961, just twenty-three days following Gagarin's flight, was greeted with great national euphoria and relief. The fear that American technology was dramatically behind the Russians was put at ease. Despite the flight's relatively short duration, Shepard was received by the nation as a hero. He was honored with ticker tape parades in Washington, New York, and Los Angeles, as well as a trip to the White House. The public outpouring of support and enthusiasm after Shepard's flight marked a dramatic and important shift in Low's perspective on the public affairs value of the space program. Until then, he had been focused too much on the congressional response and not on the public at large.

"Aside from the technical things we learned, one of the most important things I learned from Shepard's flight was that these flights should be conducted in the public eye," he said.[118] "Until that time, I had been very much against that. I thought the best thing to do was to hide behind the gates of Cape Canaveral and to conduct the flight and then tell the public about it afterwards, especially if we failed or something went wrong." His earlier opinion was most likely shaped by his NACA years of being around top secret military projects, such as the X-1 and X-15, and, perhaps, by the William Perl episode. After Shepard, he was an affirmed open-program supporter.

One of the greatest legacies of the American space program is that it was conducted as a civilian program. It operated in the open and under public media scrutiny, unlike the Soviet military's closed program. With a broad communications mandate embedded in its charter, NASA was able to convert what would have been a top secret military initiative into a global, communal experience. Having strong relations with the press ensured extensive and accurate coverage of NASA's efforts. Later, as deputy administrator at

NASA, Low would often personally court a handful of key science reporters, encouraging what he called "space buffs" to remain engaged with NASA so that they could tell NASA's story.

Through the intense media coverage, the public embraced the astronauts as national heroes. People treated them like rock stars. Low was in attendance in the ballroom at the Dolley Madison House when the original Mercury Seven were introduced to a crushing throng of news media. It was a surreal experience. He experienced the visceral outpouring of support for Shepard's flight and the attention paid at public speaking events to the astronauts who had not yet even flown a mission. The value of this star power became even more evident when Low and Mary R. accompanied Neil Armstrong, Dick Gordon, and their wives on a tour of South America during Project Gemini. In a letter to the Low children, Jan Armstrong vividly recalls the trip and the impact of the astronauts on the crowds:

> In 1966, your parents, Dick and Barbara Gordon, Neil and I all took a Department of State trip to South America for an entire month visiting the capital cities in most of the countries. The purpose of the trip was to present the Astronauts and space program to the people of South America and gain support for the US in our effort. . . . At each American Embassy in the capital city we were visiting, [we] greeted the local dignitaries at a reception. Some nights we must have received over 200 hundred local guests in long receiving lines. . . . The guests came to the event to, of course, meet the Astronauts, and, in doing so, the embassy staff were able to make personal contact with many people they had been unable to reach by other means."[119]

Like Jan Armstrong, Low was particularly impressed by the overwhelming reception of the astronauts by massive crowds.[120] He was also impressed by their ability to connect directly with the people of each country regardless of language or cultural barriers. People smiled and were honored just to be in the same room with them. The astronauts were no longer the relatively unknown military test pilots of the past. They were now the human face of American technology, progress, and freedom. They carried the heavy burden of being the leading symbols of national pride on a global scale.

In a memo to Congressman Teague, following the South American goodwill tour, Low wrote,

The press coverage of our trip far exceeded my expectations. Each paper in each city we visited devoted several pages of pictures and stories to the tour. They reported, generally with accuracy, every word that was spoken in the press conferences and each lecture that we gave. We received many questions concerning a comparison of the United States and the Soviet Union in space and found here a big plus for the United States by demonstrating time and again, that ours is an open program with results that are meant to be shared with all people and all nations. . . .

The trip as a whole gave us a wonderful opportunity to see our Latin American neighbors in the light of the United States space effort. The impact far exceeded my expectations and is perhaps a most powerful tool that the United States has in our international relations, pursued for the purpose of peace.[121]

While Low would later joke that the public was more interested in what the astronauts had for breakfast than what he had to say, there was no escaping the power of what they were experiencing to influence and impact the hearts and minds of people around the world. It was the same phenomenon that Webb or Gilruth or other senior NASA officials experienced when von Braun was in the room; he had the media power and the aura, and people gravitated toward him. This experience further solidified in his mind that space exploration was more than just a Cold War chess game; it was a major expression of national prestige and culture, a communication tool to be used in order to help establish connections between nations, and a chance to peacefully transform the technology birthed by war.

On 5 May 1961, at 9:34 a.m. EST, Alan Shepard launched aboard his Mercury Redstone rocket and became the first American in space. It was a watershed moment for NASA and for Low. "I was tremendously elated the day after Shepard's flight," Low said.[122] Later in the week, he narrated a color film presentation of the flight for Congress and the press.[123] They assembled for the occasion in the Caucus Room of the Old House Office Building, the oldest congressional office building. The film had no sound, and it was projected onto a relatively small screen. Still, the packed room marveled at Shepard's calmness during the fifteen-minute duration of his flight. They hung on every word of Low's description. "Alan Shepard's leap into space is by far the most exciting flight we have ever witnessed," he told the crowd.

When Low first heard the reports of the launch of Sputnik on his car radio, bouncing around on the dusty Ohio back roads, NASA was technically and operationally several years behind the Russians in human space launch capabilities and development. Now, they were rapidly closing the gap. But there was much more work still left to be done. Caught up in the emotional energy that Shepard's flight unleashed, the annoyances and concerns of issues like the Hornig committee report or not being first no longer mattered to him. Flying back from the Cape that evening, he got into the office just before 5:00 p.m. "Mary R. didn't even know I was in town yet," Low explained, as he went around the office to congratulate everyone who worked on the flight. "I invited everybody and anybody that I could find to a party at my home that evening." He stopped off at the liquor store on the way home, and the ensuing party raged into the early morning hours. He recalled it as probably one of the best parties they had ever thrown.

Low remembers going almost directly from the party to his office the following morning.[124] Tempered by NASA's newfound success and the national reaction to it, he couldn't wait to get back to work. With just a quarter of an hour of human spaceflight under the nation's belt, it was time to take Dryden at his word. It was time to forget about the negativity in the White House and the naysayers who said it couldn't be done. It was time to start thinking boldly. A reluctant Eisenhower administration was now out of office. There was a new, young president in the White House and a new NASA administrator, who was less averse to human spaceflight, at the helm. They were looking for a project to focus the nation's attention. It was time for Low to make his final push for a long-range goal to give the human spaceflight program a valid reason for existing—to make the risks worth the cost. NASA, Low was convinced, should focus its manned spaceflight efforts, now more than ever, on landing a man on the moon and returning him home safely. Based on his calculations, with the right amount of money and the right amount of support, it could be achieved before the end of the decade.

3

Toward a Worthy Goal

On a very humid, rainy Thursday, 25 May 1961, Low was in Tulsa, Oklahoma.[1] He was preparing to speak at the First Symposium of Peaceful Uses of Space, which was set to kick off formal presentations and panel discussions that Friday and Saturday. The symposium, sponsored by NASA and the Tulsa Chamber of Commerce, would be followed by a two-week space fair. Open to the public, the fair was a PR push designed to showcase the nascent industry's new technology and engineering prowess. It was an anxious time for NASA management. NASA had not yet received any direction from the newly installed Kennedy administration on any follow-on human space exploration program after Mercury. And while discussions on future plans and options with Vice President Johnson were increasing, the near-term outlook was still opaque, at best. In many ways, U.S. manned spaceflight was still very much a start-up program in search of funding and a mission.

President Kennedy's science advisor, MIT's Jerome Wiesner, had also been a part of Eisenhower's science advisory committee, along with Dr. Hornig. Wiesner had issued a relatively lukewarm report earlier in the year criticizing NASA's focus on human spaceflight versus robotic programs, and he cautioned against the agency's ability to catch up to the Russians in any meaningful way. Worse, with several major launchpad explosions under its belt, NASA's focus on human spaceflight caught a hard rebuke from Wiesner on the political risks of losing an astronaut in space: "A failure in our first attempt to place a man into orbit, resulting in the death of an astronaut, would create a situation of serious national embarrassment. An even more serious situation would result if we fail to safely recover a man from orbit."[2]

On top of that, Harvard professor Richard Neustadt was skeptical of the ability of the United States to catch up to the Russians, let alone exceed them in space. One of Kennedy's top advisors, Neustadt was advocating that Kennedy divert NASA funding to more tangible military, scientific, and sociological programs. Given the growing societal and global geopolitical problems facing the administration, it was not an outlandish request. Funding of manned spaceflight truly hung in the balance on a very thin line between competing forces and interests within the White House. In *John F. Kennedy and the Race to the Moon*, John Logsdon points out that Kennedy was never all that enamored with spaceflight. Like Eisenhower, Kennedy was skeptical of the massive costs involved.

In *The Decision to Go to the Moon*, Logsdon explains further, "When John F. Kennedy assumed the duties of the presidency on January 20, 1961, he seemed to know less and to be less interested in issues of space policy than almost any other set of policy questions. . . . Until Kennedy became convinced that space achievement was linked closely to the power relationship between East and West, and was a symbolic manifestation of national determination and vitality, his lack of knowledge about space matters made him hesitate to make basic changes in the space policy developed by Dwight Eisenhower."[3]

In the shadow of this murky political uncertainty, Low and Gilruth traveled to Tulsa. By this time, Low had been pushing behind the scenes for several years within NASA for a manned lunar landing—so much so that one journalist referred to him as a "moon zealot," much to his own chagrin. At the Tulsa conference, their goal was to capitalize on the public interest sparked by the success of the recent Mercury flights. He was eager to explain the agency's future (albeit speculative) plans for Apollo, which he had recently authored.

The theme of the conference would be a political framing event in light of the lack of clear policy direction out of Washington. In a little advance PR to promote the event, Webb had published a widely syndicated article titled "The Space Age Is Changing Your Life on Earth."[4] In it, Webb argued for the benefits the average taxpayer would receive from space activities in return for the program's risks and costs. Benefits would include new products for the home, improvements in global communications and television broadcast technology, meteorological satellites, and medical instrumentation. Webb even forecasted the development of "a new type of camera" perfected for space vehicles that promised to someday produce movies and instant still pictures without chemical processing. Even before Apollo was officially sanctioned, NASA began emphasizing space-exploration spin-offs in an attempt to curry public favor. Still, for most people, it all came across as science fiction.

Given the political headwinds, it is hard to overstate the importance of the success of Alan Shepard's short, suborbital ballistic flight. Low was convinced that the ensuing public euphoria temporarily eased many of the Kennedy administration's concerns. After the disastrous Bay of Pigs incident in Cuba, Kennedy needed an audacious policy win—not just for his own political ambition but also for the country's standing on the international stage. Kennedy wanted to find something—*anything*—that the United States could do in space that would put it ahead of the Russians. He tasked Vice President Johnson to survey NASA, the military, and industry to come up with a suitable and technically feasible goal. Johnson, a longtime advocate of space exploration, had served on the Senate Special Committee on Aeronautics and Space. He had recently been appointed by Kennedy as chairman of the newly formed National Space Council, charged with overseeing all of NASA's activities. Johnson solicited various opinions by turning to a broad cross section of government and industry representatives—including NASA administrator Webb and deputy administrators Dryden and Seamans. Webb and Seamans drew heavily on a series of lunar landing program studies, plans, and projections that Low and his team had recently completed. Still, no formal decisions—as far as Low or anyone else in NASA management knew at the time—had been made.

During the middle of the Tulsa conference, however, the country's first television-era president shocked the assembled crowd, and, indeed, the nation, by suddenly making space a central theme in a speech to a special joint session of Congress. Kennedy's historic, nine-point speech was broadcast live to conference attendees.[5] Echoing Low's caution against being first in space, Kennedy echoed Low's reasoning and support for a lunar landing when he made his famous pitch. It was as if Low had written it himself:

> Now it is time to take longer strides—time for a great new American enterprise—time for this nation to take a clearly leading role in space achievement....
>
> For while we cannot guarantee that we shall one day be first, we can guarantee that any failure to make this effort will make us last. We take an additional risk by making it in full view of the world, but as shown by the feat of astronaut Shepard, this very risk enhances our stature when we are successful. But this is not merely a race. Space is open to us now;

and our eagerness to share its meaning is not governed by the efforts of others. We go into space because whatever mankind must undertake, free men must fully share. . . .

I believe that this nation should commit itself to achieving the goal, before this decade is out, of landing a man on the moon and returning him safely to the earth. No single space project in this period will be more impressive to mankind, or more important for the long-range exploration of space; and none will be so difficult or expensive to accomplish.[6]

Upon hearing Kennedy endorse a lunar landing program, Low was thrilled. After all, he had been pushing for Apollo, and pushing specifically for a lunar landing, since 1959. "The overall impetus to make Apollo a lunar landing program came primarily from me in Washington," Low explained, describing the moment. "Several key facts came together at the right time. Through the winter of 1960–61, we completed a large number of studies [the so-called Low committee report] that showed a lunar landing would be feasible in the decade. The success of Al Shepard's flight, the phenomenal public interest it spurred, the availability of a NASA program to land on the moon, and the administration's need to have something other than the Cuban situation to focus on. All of these things, together, made the time right to decide to go to the moon."[7] It was a rare, brief, geopolitical moment in time, and NASA was ready to seize it primarily because of the work Low had done.

While Low was enthusiastic, Gilruth was shocked at the audacity of the goal. "When I heard President Kennedy announce that American spacemen would land on the moon, I could hardly believe my ears," Gilruth admits. "I was literally aghast at the size of the project."[8] This was understandable, Low explained. Like the majority of the STG at the time, Gilruth was focused almost exclusively on conducting the Mercury program. So far, NASA had only conducted theoretical planning and industrial studies for a lunar mission under Low's guidance. For all intents and purposes, it had been just a drawing board exercise, a bit of extracurricular activity. Now, landing a man on the moon before the end of the decade was officially U.S. national policy. "Bob was looking at what was still left to be done in Mercury, and he felt going to the moon was too ambitious of a step at that time," Low explained. And Gilruth was right. Gilruth's concern would later be the leading cause behind his own passionate support for the Gemini program. Project Gemini, conducted between 1961 and 1966, was designed to develop and demonstrate the critical space travel techniques—things like working outside the spacecraft and tracking and docking in space—that would be needed to support Apollo. "Bob knew we needed something operational between Mercury and Apollo to demonstrate space operations," Low said. "At our planning retreats on Wallops Island, even before Shepard flew, Bob was focused on having an advanced Mercury program. The big item of discussion at the time was 'Where do we go from here?'"[9] The "Where do we go from here?" discussions would continuously come up between Low and his team and NASA management in Washington, for planning purposes, and between all of them and Gilruth and his operations people at Langley, for actual mission planning.

Low recalled that there were several motivational factors to having a program between Mercury and Apollo.[10] First, there was a need to figure out maneuverability in space—in terms of both the spacecraft and the astronaut outside the spacecraft, let alone the critical ability to rendezvous between two spacecraft. Second, there was a need to develop and test a modular approach to the spacecraft design, to deal with the increasing complexity of the machines and the mission requirements. "Another major motivation was how difficult Mercury was beginning to be with checkout," Low said, detailing the technical problems that would arise as the missions and thus the spacecraft became more complex. "Every time we had to check one out after receiving it from the contractor, we had to take everything back out of the spacecraft and rebuild it down at the Cape. It was a real problem." It was about this time, Low recalled, that Jim Chamberlin, on Gilruth's Langley team, began advocating for a modular design in the Gemini spacecraft—putting critical systems on the outside of the spacecraft instead of on the inside. Repairs and tests could be done more quickly if you didn't have to take the whole thing apart just to get at the problem. Chamberlin also advocated strongly for a two-person crew, especially in light of extended, multiday missions as planned for Apollo. "There's no question in my mind that Chamberlin is the designer and inventor of Gemini," Low said. "But without Bob immediately latching on to that idea and pushing it, we wouldn't have had a Gemini. He believed that we needed to learn how to fly in space; we needed to learn to do more than we were able to do in Mercury before we can land a man on the moon." Gilruth remained convinced that without Gemini, the goal to accomplish the first landing by the end of the decade would never be achieved.

Still, despite the obvious need for an operational program like Gemini, it did not have a lot of strong support in NASA management back in Washington, with the exception of perhaps Low, and he had to push to keep it alive. Management at the time—including Brainard Holmes, who took over after Silverstein left in the fall of 1961—was more focused on Apollo after Kennedy's speech. They were making the political leap to the next program, instead of the logical technical step to the next level of capability. "Holmes even said that if he had not been that new, he would not have approved it," Low said, in way of clarifying the lack of support within NASA Headquarters for Gemini. "He said this during Gemini's darkest days. The approval of the program was a result of the tremendous desire to do it by Gilruth. There were very few people throughout Gemini that had the perseverance and the wisdom that Bob had in pushing it through and keeping it going. It was his program."[11] So despite his initial hesitation in Tulsa, Gilruth was as committed to the lunar goal as Low.

With a large, national-media spotlight now on Low's planned symposium speech, he introduced the country to the current planning state of Apollo. With Kennedy's speech grabbing the attention of the nation, quotes from his presentation were carried by all the major broadcast and news wire services. According to the *New York Times*, Low "gave as detailed a description as has yet been made public of the Apollo space craft which is expected ultimately to land an American on the moon."[12] In addition to describing what would become the Apollo spacecraft command and service modules, Low explained how the moon landing would be gradually attempted. First, he told the rapt audience, Apollo vehicles would be flown into Earth orbit; then, the orbits would gradually be expanded, including voyages around the moon and ultimately to a moon landing and safe return. It was broad and generic without a lot of detail, but it was like chum in the water for the assembled press corps. They ate it up. "This is the first time that a president has given us the go ahead for a lunar landing," he told the United Press International.[13] To the Associated Press, in a quote carried by newspapers around the nation, he discussed the exciting scientific questions that lunar exploration might one day answer: "Primordial life forms *may* lie preserved in the dust which has gathered on the moon over the last four billion years; moon dust samples one day *will be* brought back to Earth."[14]

***

So how did it come to pass that George Low happened to be so focused on a lunar landing at the right place and at the right time? Historian Henry Dethloff, writing in *Suddenly, Tomorrow Came*, remarks that to the outside world, Kennedy's lunar landing mandate seemed like it "was thrust upon NASA," but in reality, "it was a project invited, planned for, dreamed of, and enthusiastically entered into by the NASA community."[15] NASA employed no one more enthusiastic and creative around such a dream as Low. It can be argued that no one could claim more credit in getting the goal across the finish line with a presidential mandate than him, either. In the earliest days of NASA's formation and for a full two years before Kennedy's speech, Low was appointed to the Goett committee as Abe Silverstein's representative. John M. Logsdon, in *The Decision to Go to the Moon*, describes the formation and goals of this important committee, as well as Low's central role in being one of the first to advocate for it:

> Since there was no specific objective for post-Mercury manned space flight, the first task of NASA was to choose one. To do this, NASA formed, in April of 1959, a Research Steering Committee on Manned Space Flight. This committee, chaired by Harry Goett of the Ames Laboratory, came to be called the Goett Committee. . . . The committee proceeded by reviewing existing manned flight and launch vehicle programs, and the programs of research related to manned space flight then in progress in NASA centers. George Low of the Office of Space Flight Programs in NASA Headquarters recommended to the committee that a manned lunar landing mission should be adopted as the long-range objective of the NASA manned space flight program.[16]

At the meeting, other members felt that a more moderate approach would be an easier task to achieve.[17] They suggested a manned flight around the moon, or a circumlunar flight, rather than actually landing. Not willing to give up on the concept of a landing, Low tried another approach. Cognizant of what would be more acceptable to Congress, let alone to a budget-conscious White

House, he suggested the committee adopt a lunar landing mission not as a *present* goal but as a *long-range* objective. With proper emphasis on intermediate and long-range steps, he was sure such an iterative approach would be easier to sell and eventually get funded.

In addition, he urged using existing launch vehicles—such as the army's Saturn booster family of rockets currently in development by von Braun's team in Huntsville—instead of undertaking the costly and time-consuming effort to develop new and larger launch vehicles, like the proposed Nova rocket.[18] While the Goett committee did not come to agreement in this first meeting, Low continued to lobby the members for his ideas. Other concepts were introduced and explored. These concepts included launching an Earth-orbital space station or undertaking interplanetary travel beyond the moon to Mars. Mars, he knew, would be too expensive to attempt. As the team talked and Low lobbied, they eventually coalesced around Low's long-range, intermediate-step approach. This approach, he reasoned, would help the nation develop a more sophisticated spacefaring capability.

"As I recall, the committee deliberated whether the next major step in manned spaceflight should concern itself with a large space station or a deep-space program," said Low. He further detailed their deliberation:

> The basic conclusion was that a deep-space program—and by deep space, I mean anything beyond earth-orbital activities—would lead to a quicker, focused advancement of the technology needed for manned spaceflights. We decided it should be a lunar flight, because it was a more difficult job to do. As a committee, we also decided to stop just short of a landing. The committee felt a landing was a step which was too far beyond what NASA was willing to talk about at that time. In fact, I remember Harry Goett being asked in one of the meetings, "When should we decide whether or not to land on the moon? And how will we land on the moon?" And Harry said, "Well, by that time, I'll be retired, and I won't have to worry about it."[19]

Low, of course, felt otherwise, but he went along with the consensus for the time being, especially if the landing was included as a longer-term goal.

In the final *Long-Range Plan for Manned Space Flight* delivered in December of 1959, the committee gave a strong nod to Low's approach. They put a tentative pin in the calendar for a lunar landing at "some-time beyond 1970." This would take place after a program leading to human circumlunar flights and a permanent near-Earth space station for the 1965 to 1967 time frame. It was also a program more consistent with the von Braun paradigm. With a time line and a mission-goal framework in hand, NASA began some preliminary studies necessary to advance future circumlunar flights. Meanwhile, von Braun's rocket team had been transferred to NASA control from the army, and Eisenhower instructed Glennan to accelerate what he called "the super booster program," which would lead to the Saturn family of rockets.[20]

The advanced-spacecraft design team of the STG began working on a modular system, with different staged functions for various phases of a lunar mission. This module approach would eventually lead to the Apollo command, service, and lunar modules. Other NASA centers began studying critical flight program phases, from advanced guidance systems to biomedical considerations, to actual flight and maneuverability concepts. One such concept was a "soft rendezvous," or how two vehicles could come together at the high velocities required of space travel without crashing into each other. This idea was being passionately researched and advanced by John C. Houbolt at Langley, and it would prove to be critical to the future success of Apollo. Without Houbolt's work and his dedicated attempts to get it noticed by NASA management, a lunar landing most likely never would have been possible.

As the various committees and NASA center research groups did what they could amid the actual operational challenges of Project Mercury, Low became convinced that it was time to reach out to private industry. Internal analysis was one thing; it was another to get the aerospace community involved and making actual proposals on equipment and options. Low worked with Silverstein to put together a presentation for Glennan, Dryden, and Horner—the top three executives in NASA management at the time.[21] The goal was to get their formal permission to host a NASA industry conference in Washington, as well as a million dollars to commission two or three industry feasibility studies—no small ask from an Eisenhower administration weary of human spaceflight.

"The presentation was designed to give them a description of the advanced manned spaceflight programs as we then saw it, including its scheduling implications and its cost," Low explained. The meeting took place on Saturday morning, 9 July 1960, in the old NASA Headquarters building on H Street, just kitty-corner to Lafayette Square and in sight of the White House. Initially, Silverstein was scheduled to make the formal presentation and request. But he had been traveling out of town. Low prepared some preliminary slides, and the two began to review them late on Friday night, right before the meeting. They worked on the slides and the presentation well into the evening. As Low schooled Silverstein on the details and background on Apollo, somewhere around midnight, Silverstein suddenly stopped. He had heard enough. The true expert on the subject was sitting across from him. Why should he try to parrot what Low already had in his head and perhaps risk losing the argument? He turned to Low, pushing the slides toward him.

"You make the presentation," he said.

Low was up the rest of the night, finalizing his charts, practicing the flow of his thoughts, going over every aspect down to the smallest detail. The next day, in the executive conference room, he made a flawless and impassioned presentation. "George came in unshaved, no necktie," recalled Glennan with amusement. "He had been up all night, and he presented the concept of Apollo to us as it would later fly. It was an amazing performance."[22]

It was also at this meeting that Low introduced Silverstein's program name: Apollo.[23] Silverstein, who had named Project Mercury, wanted to create a tradition of naming NASA projects after Greek gods. He picked Apollo out of a Greek mythology book because, in his mind, the image of Apollo riding his golden chariot across the sky represented the grand scale of the proposed program. Low remembers, after Glennan approved the name, writing in his notes, "And we will call this program, 'Apollo.'"[24]

On the strength of Low's presentation, Glennan gave them the go-ahead for the industry conference. He also gave permission to fund the three industry research studies out of their already stretched budget, so there would be no need to go back to Congress to ask for new spending appropriations. Glennan reminded Low, however, that these were to be research studies *only*. There was no formal White House approval for any program past Mercury. Low was pleased. Should NASA ever get White House approval or should the mood in the nation or in Congress change suddenly in favor of a more aggressive lunar program—as it would after Kennedy became president—they would at least be prepared. Even more remarkable, Low was able to sell the idea to Glennan at a time when Mercury was still struggling with a 62 percent test-launch failure rate. NASA had not yet even launched an animal, let alone a human, into space. Their rockets were blowing up six times out of ten. Now, they were funding industry studies to go to the moon.

It was as forward thinking and bold as anything on the table in Washington at the time.

On Thursday, 28 July 1960, just four months before the November elections, over 1,300 representatives from government, the aerospace industry, and educational institutions traveled to Washington to attended NASA's first industry planning conference.[25] The press caught wind of it, and they were in attendance as well. The conference involved over twenty NASA program officials outlining the agency's various plans for launch vehicle development and potential projects for both manned and unmanned spacecraft and satellites. As scheduled, Low announced and presented the proposed Apollo lunar program.

Press reports out of the meetings the next day excitedly announced NASA's bold plan for a "vast space program" to cost between $12 and $15 billion, including both human and robotic space missions.[26] NASA's program would be built around four key areas: (1) Project Apollo, intended to carry three astronauts into lunar orbit sometime after the first Mercury orbital mission; (2) Project Aeros, a geostationary satellite set to be placed in orbit at the earth's equator and traveling in tandem with the earth's rotation; (3) Projects Prospector and Surveyor, for robotic lunar scientific exploration; and (4) a series of passive communications satellites for bouncing radio and television signals around the world. It was a big, bold series of projects, but Apollo dominated the media's attention. Unfortunately, not a word was mentioned that Apollo was only in the study phase. It was presented as if it was a done deal. NASA was planning to go to the moon.

"The general subject of the NASA industry conference was to describe industry studies we would soon request for a circumlunar flight program," Low said. "But at the same time, we also informed industry that we had no approval for the program beyond these studies. We were talking only about a study effort."[27] Naturally, those cautions fell on deaf ears with the press, as they, too, were caught up in the excitement of a lunar mission. From a timing perspective, he explained, "Apollo was a program with two avenues of approach. One of them, the mainstream, would be a circumlunar flight. We had planned a program that, within this decade, would lead to a circumlunar flight. Beyond the decade, I told the conference that Apollo would eventually lead to a lunar landing and to planetary exploration. I also said that, within this decade, Apollo *could* lead to, or *could be part of,* the space station program."[28]

He presented what he described as a rational, reasonable approach to a long-range development program leading to the manned exploration of outer space. "The Apollo vehicle, to be launched by Saturn rockets, will be flexible enough to be capable of both flying around the moon—an intermediate step toward future manned landings on the moon—and serving as a manned laboratory," Low told the group.

> Apollo will be built on the modular or building block concept, with three basic parts: a command center, which would house the crew during the launching and during re-entry to the earth's atmosphere; a propulsion unit, which would provide course corrections, place the craft into an orbit around the moon, eject it from orbit, or maneuver it during earth orbits; and a mission unit, which could be changed for various flight missions. The Apollo spacecraft would represent an intermediate step toward establishment of a permanent manned space station above the earth, which could then lead ultimately toward manned landings on the moon and planets.[29]

Concurrent with the start of the conference, and unknown to most in attendance early that morning, the first Mercury capsule (MR-1) launch attempt, aboard an Atlas rocket at 9:13 a.m., exploded in spectacular fashion just sixty-five seconds after takeoff from Cape Canaveral. NASA's launch failure rate still exceeded its successes. "The first working model of America's man-into-space capsule apparently was blown to bits miles high in the sky today," the United Press International reported.[30] NASA would not announce the news until several hours later, around noon eastern standard time and only with a short, terse statement. Morale within the STG was at its nadir. The test flight warranted only a few short lines in most newspapers, drowned out by Low's Apollo presentation. The unintended consequence of Low's enthusiasm, of course, was the negative interpretation of the press reports by the Eisenhower administration.

At the end of the industry conference, potential bidders were requested to attend follow-up sessions in the coming weeks. These sessions would spell out mission specifics, such as a "shirt-sleeves environment" in the Apollo capsule, radiation-protection requirements, atmospheric control, and other similarly granular details. On 12 September 1960, the STG published its formal Apollo request for proposal.[31] Eighty-eight firms attended the bidders briefing; sixty-six picked up printed submission forms; and by 9 October 1960, NASA received fourteen competing bids. By the end of the month, NASA would award three $250,000 study contracts. The firms selected would go on to spend considerably more than they were paid: Convair spent over $1 million; General Electric, over $2 million; and the Martin Company, which sent in the most elaborate and detailed study of the three, put over three hundred people on the project for six months, spending over $3 million on the data and designs for their feasibility recommendations. Martin provided over a dozen spacecraft configuration models and a final report of over nine thousand pages. According to *Chariots for Apollo*, the Martin proposal focused on "versatility, flexibility, safety margins, and growth," and it was the only study to detail "the progression of steps from a lunar orbiting mission to a lunar landing." Despite Martin's extra effort, the overall goal of the feasibility studies, and the emphasis of the mandate, was still a circumlunar and not a lunar landing mission. The industry studies, nonetheless, were well worth the investment. At a cost of just $750,000 to taxpayers, private industry returned to NASA over $6 million worth of advanced lunar mission planning and research studies—an eightfold return on their small budgetary investment.

It was also a large return for a program that so far officially did not exist. In fact, no employees were even allocated to it. "The only one in Washington who really spent any time on advanced programs was John Disher, and he probably didn't spend more than 10 percent of his time on it, because he was also involved with the Mercury program," Low said. "Apollo was all still very much an extracurricular activity."[32]

During the fall of 1960, a number of additional lunar landing studies were conducted. "For the first time, we began looking at what it might take to make the circumlunar mission into a lunar landing mission," Low said.[33] As the year progressed and studies continued, Seamans wanted to get a sense from Low about where things stood with lunar mission planning, especially as it related to booster capability. Seamans called a meeting of the so-called Space Exploration Program Council (SEPC) of NASA. SEPC was a group established by Glennan earlier in the year to oversee NASA program planning and implementation. The group consisted of Goett, von Braun, William H. Pickering,

Ira H. Abbott, Silverstein, General Ostrander, Albert Siepert, and Seamans. The meeting was held on Friday, 30 September 1960, and Low's dramatic, Socratic-style presentation to the SEPC is chronicled in *Chariots for Apollo*:

> Low posed five questions and defended his answers to them as proof of the realism of the proposed schedule for Apollo: (1) Will the spacecraft be ready on time to meet the Saturn schedule? (2) Will the spacecraft weight be within Saturn capabilities? (3) Are there any foreseeable technological roadblocks? (4) Will solar flare radiation prevent circumlunar flights by men? (5) What are the costs for this program? To each of these five questions, Low made positive assertions of competence and capability. He argued that an Apollo circumlunar prototype spacecraft could be ready in three to four years, a production vehicle in twice that time. Space Task Group weight estimates showed a reasonable margin between the weight of the spacecraft and the payload the C-2 Saturn could be expected to boost. No insurmountable technological obstacles were anticipated, Low said, not even reentry heating or solar flare radiation. Low concluded that the current cost level of $100 million a year would eventually rise to approximately $400 million annually. All of these considerations, in [Low's] opinion, argued for an immediate decision to go ahead.[34]

The media spotlight generated by the NASA industry conference, Low's quotes in the press reports, and the awarding of the feasibility studies attracted the attention of the White House. This attention led to Hornig's committee investigation into NASA's activities. "As far as we can tell," the committee reported, "the (overall) NASA program is well thought through, and we believe that the mission, schedules and cost are as realistic as possible at this time." As for Apollo, "any of the routes to land a man on the moon [will] require a development much more ambitious than the present Saturn program.... Nevertheless ... the first really big achievement of the man-in-space program would be the lunar landing." The panel informed Eisenhower that building a booster large enough to facilitate a lunar program would be a budget buster, costing anywhere between $25 billion and $38 billion. In response to the Hornig report, the headlines Low generated out of the industry conference, and the large price tag, Eisenhower drafted a budget that specifically stated there would be no manned program after Mercury.[35] It took Glennan and Dryden both to calm him down and persuade the president to back away

from such a hard close. They asked him to amend the budget with a softer statement indicating that more research was required before the ultimate course
of manned spaceflight after Mercury could be determined.

Despite having his plans almost push the human space program to extinction, Low was undeterred. Seeing how close NASA was to losing a human
spaceflight program, he became convinced that the agency needed to make a
stronger case. While the *how-to* studies for the circumlunar flight were ongoing, he felt that the team needed to spend more time on the *justifications* for a
human lunar landing program. "I was quite concerned, of course, that in the
subsequent year's budget, there were insufficient funds for any major lunar
program," Low admitted. "I felt it would be important to have something
in the files. We needed to be prepared to move out with a bigger program,
should there be a sudden change of heart within the government, within the
administration."[36] Once again, he took it on himself to rectify the situation,
pulling together a small, ad hoc working group. To formalize and legitimize
this working group within NASA management, he wrote a short but seminal
memo to Silverstein on 17 October 1960:

> It has become increasingly apparent that a preliminary program for
> manned lunar landings should be formulated. This is necessary in order
> to provide a proper justification for Apollo, and to place Apollo sched-
> ules and technical plans on a firmer foundation.
>
> In order to prepare such a program, I have formed a small working
> group consisting of Eldon Hall, Oran Nicks, John Disher, and myself.
> This group will endeavor to establish ground rules for manned lunar
> landing missions, to determine reasonable spacecraft weights, to spec-
> ify launch vehicle requirements, and launch vehicles. This plan should
> include a time phasing and funding picture and should identify areas
> requiring early studies by field organizations.[37]

In one of human spaceflight history's greatest, most understated moments,
Silverstein responded to Low by writing his approval in pencil in the memo's
margin:

Low,

O.K.

Abe.

With approval in hand, Low set out to work. The team, later known as the Low committee, was eventually formalized by Seamans into the Manned Lunar Landing Task Group, in early January 1961.[38] Low was appointed chairman. Seamans instructed the committee to prepare a position paper for NASA's fiscal year 1962 budget proposal. John Logsdon describes the significance of this report:

> The paper was to be a concise statement of NASA's lunar program . . . and was to present the lunar mission in terms of both direct ascent and rendezvous. The rendezvous program would be designed to develop a manned spacecraft capability in near space, regardless of whether such a technique would be needed for a manned lunar landing. In addition to answering such questions as the reason for not eliminating one of the two mission approaches, the group was to estimate the cost of the lunar mission and the date of its accomplishment, though not in specific terms. Although the decision to land men on the moon had not been approved, it was to be stressed that the development of the scientific and technical capability for a manned lunar landing was a prime NASA goal.[39]

In just shy of a month's work, the committee submitted its report on 7 February 1961. In the final analysis, the team felt that no technical invention or breakthrough would be required to insure the overall feasibility of a safe lunar landing.[40] A lunar landing via either lunar-orbit rendezvous (LOR) or direct ascent—that debate had not yet been settled, despite Houbolt's continued attempts to push for an LOR—could be accomplished by the end of the decade.

Subsequently, Seamans and Dryden used the Low committee report as the basis for discussing and advocating a lunar landing mission with Webb. At the time, Webb was reviewing the Eisenhower budget for NASA and working on proposing a new budget to Kennedy, via Johnson. The difference between Glennan and Webb in terms of supporting human spaceflight was like night and day. For his part, Webb drew heavily on personal input from Low. "In 1961, when there was a question about the continuation of a manned spaceflight program beyond Mercury, it was George Low that was chosen by Dr. Dryden to brief me on the importance of manned spaceflight," Webb explained. "He convinced me to the extent that I made three speeches within the first weeks serving as administrator pledging that we would go forward with manned spaceflight."[41] Despite a contingent of NASA scientists favoring robotic missions over human spaceflight and pushback from the outgoing Eisenhower administration, the main consensus now within NASA favored human space exploration.[42]

"The report by George Low and his group was most valuable in the meetings with the president and Congress that were soon to follow," Seamans declared.[43] Webb, Dryden, and Seamans were united around Low's conclusion when Johnson came calling for a new, ambitious goal for NASA. "Jim Webb thought so much of Low's leadership ability, he tapped him to chair the agency committee that placed on President Kennedy's desk the report recommending a manned lunar landing," explained Alan Shepard. "Modest and unassuming, he was a methodical man who sought harmony and order. Low brought Kennedy to the podium to commit America to the challenge of the century."[44]

As Low had anticipated, a rare window in geopolitical need and time had, indeed, opened. The window was short, and it was one of those unique moments in human history when a bold idea had probably just one shot at getting through or would risk being lost forever. The fundamental answers Low addressed with his committee—including cost, timing, and modes of landing—formed the basis for Johnson to recommend to Kennedy a manned lunar landing before the end of the decade. "When President Kennedy's White House called up Jim Webb and said, 'Now, what about this moon mission?' NASA already had the answers, primarily through the work of the Low committee," confirmed Faget.[45]

Kennedy, apparently, agreed.

After Kennedy's speech, the nation now had a goal in manned spaceflight worthy of the expense, risk, and effort. It was *the* goal that Low had envisioned for NASA back in 1959 but that Eisenhower, his advisors, and even Glennan refused to fully embrace. For two years, Low lobbied, pushed, cajoled, and, in the end, through his own sheer force of will, helped get sanctioned. Nonetheless, with the Low committee report being the primary impetus behind advocating for a lunar landing, Kennedy boldly moved Apollo out of the purely theoretical and into the realm of operational priority.

Now, the really hard work was about to begin.

With practically all NASA's manned spaceflight personnel focused exclusively on preparing for only the second manned Mercury flight, NASA had to figure out the technical and engineering aspects of a lunar flight, not to mention the organizational and production capabilities to achieve such an audacious goal. It was no small order. It had taken the fledgling agency almost three years just to achieve fifteen minutes of human spaceflight time. Kennedy was giving them just nine years to land a man on the moon and bring him safely back to Earth. There was no time to lose. "George Low's contribution to Apollo must rank above all his others," Silverstein proclaimed, looking back at this particular period in U.S. spaceflight history. "All that went before was prelude."[46]

***

After Kennedy embraced the lunar landing goal, Low continued on in his administrative role at NASA Headquarters under Silverstein. He helped the agency finish out Mercury, shuttling back and forth from the Cape to NASA Headquarters, and participated in major management decisions, such as figuring out the practicalities of how to actually land a man on the moon. From helping to advance the LOR decision, through to the changes in organizational structure within NASA as it grew, he continued to find engagement and satisfaction in Washington.

Still, the promise that Silverstein made to him so many years ago—that he would one day be able to work in a field center—tugged at him. The dirty-hands engineer of his youth, the one inside him who liked to take machines apart just to see how they tick, had never really been extinguished in his heart. He yearned to get back into the field. This desire resurfaced when Silverstein himself left Washington to return to Cleveland shortly after Kennedy's pronouncement. "At Abe's farewell party we had for him, he said, 'You know, George Low always said the field is the place to work. I'm going back, even though he isn't,'" Low recalled, a bit bemused.[47]

To replace Silverstein, Webb hired Brainerd Holmes, an electrical engineer with a long pedigree of experience at Bell Labs, Western Electric, and most recently RCA, working in missile development and military development projects. Sensing an opportunity and cognizant of Low's prior wishes, Gilruth once again asked Low to join his team in Houston. After all, he had been a member of the team for a few short weeks back in 1959, before he was summoned back to Washington.

"At the time, he offered me the job of being the number three man to him and Walt Williams. I told him that I wanted to meet Brainerd Holmes first," Low said.[48] In reality, the timing just wasn't right. He didn't want to leave his current post just to be a number three. He was also aware that several of Gilruth's deputies were trying to jockey for position and "push Gilruth aside."[49] He didn't want any part of that. "Gilruth went after strong people," Low said, underscoring the ambition that is usually associated with type A, strong-willed people. Low didn't think of them as bad people; many, he said, accomplished great things. It just wasn't the environment that he liked or wanted to work in. His operating model centered on the team. After balancing the meeting with Holmes against Gilruth's offer, he quickly made the right decision for himself and the time. Plus, he really hit it off with Holmes. "I felt he needed me, and he could use me," said Low, who felt it would be important to the program to provide continuity and the support that Holmes would need as an outsider transitioning into NASA.

Though Holmes would only be in the job for two years, his impact was nonetheless important—on NASA as well as on Low. While NASA was desperately trying to catch up in technical progress with the Soviets, the Apollo program was falling further behind in both scheduling and budget. Key decisions were mired within interagency disagreements and power struggles. It was difficult to bring together the leadership of the various NASA centers. The management problems he had experienced in the early days of Mercury were now compounded by the massive crash-course program being pushed to achieve Kennedy's end-of-the-decade goal. "I found myself in Washington with an illustrious title, an organization of somewhat conflicting interests, and the challenge of establishing a program to send a man to the moon," Holmes once said in describing the situation. "It was a formidable assignment."[50]

With Low's help, Holmes streamlined the agency's management structure and improved internal communications and cooperation through formal management councils across the key NASA facilities. More importantly, he finally approved LOR as the mode for landing on the moon. Even though he was initially against LOR, Holmes was won over by the arguments made in the management council meetings. Gilruth would later credit Holmes's management-team approach, especially at this critical time in development, with being a key element of the Apollo program's ultimate success. In the Holmes-chaired meetings, Gilruth observed, management "argued out" various opinions before setting course. It was a key method of getting everyone on the same page, especially among the independent-minded leadership of NASA's various centers. "A less skilled leader than Holmes might have forced an early, arbitrary decision," Gilruth explained. "That would have made the whole task of getting to the moon virtually impossible."[51] Seeing the positive effect of the council in moving decisions along and getting everyone on the same page was also not lost on Low. "Brainerd was a different sort of guy than Abe," Low said, summing up the simple but effective approach. "He knew management, where Silverstein had known the technical things. I thought he did an outstanding job for the organization."[52]

Of course, sometimes happenstance also plays a role in decision-making. Low tells a rather humorous story about Holmes's first management council meeting and finding out that von Braun wanted the Saturn V rocket's first stage to have five engines, instead of the original four as designed. The meeting was in Holmes's office at the corner of Nineteenth and Pennsylvania in Washington. He had a contractor model of the rocket that would eventually become the Saturn first stage. "It had four engines in it," Low remembers, describing the scene. "Von Braun and a whole number of us were standing around the table before we went into the meeting, looking at it. And von Braun starts saying, 'This is where we want to put the other engine.'" Low was surprised. This was the first he had heard about a fifth engine. "Why do you want another engine," he asked von Braun. "Well, there's this big hole," von Braun said, a bit of a smile breaking out as he motioned to the open area in the middle of the four engine bells, "that is just crying out for another engine."[53]

In recalling the story, he wasn't sure of von Braun's exact words or if he was just paraphrasing. "But he did, in I think a very forceful way, say this is a capability or capacity that this vehicle should have, and we ought to put it in." Low said. "I remember talking to Bob on the side and asking him what he thought of that added complication of one more engine. And Bob said, 'Look, I'm so dedicated to going by LOR, I'm sure we'll be able to use the added weight-lifting capability if we had the fifth engine. So, let's go ahead with it.'" And go ahead with it, they did. The Saturn rocket first stage would eventually have five Rocketdyne F-1 engines—four outer engines that could be turned, or gimballed, and one stationary one right smack in the center, as von Braun wanted "to fill that hole that was just crying out for an engine." During the

Apollo program a total of fifteen flight-capable Saturn V rockets would be built, but only thirteen would be flown. For more than a half century, it has been the only launch vehicle to ever carry humans beyond low Earth orbit.

During this period, Gilruth again tried several more times to poach Low, but he always declined. The management structure down in Houston had not yet changed to Low's liking, and he was making important progress with Holmes in Washington. Holmes also hired several other key figures, including Joe Shea, as a deputy for systems. Shea, a charismatic Irishman with a crew cut, soft voice, and a penchant for wearing red socks, had worked for AC Delco, Bell Labs, and TRW before joining NASA. He would figure prominently in the buildup of Apollo from this point until the *Apollo 1* fire. Low took an almost immediate liking to Shea, and he felt that Shea was technically a very strong engineer. "As a team, we had fun, and we were doing good work," Low explained. So what would the point be in leaving? He was shocked, then, when Bob Seamans called to inform him that Holmes was leaving. It was the result of friction with Webb—over budgets and the autonomy Holmes wanted in running the office as he saw fit. Holmes insisted; Webb refused; Holmes made an ultimatum; Webb showed him the door. "I was quite shocked. I had been loyal to Holmes. I had been close to him," Low said.[54]

To replace Holmes, Webb hired another outside industry veteran, George Mueller, who came to NASA from TRW, where he had worked with NASA and the military on various missile and radar projects. Mueller and his team were already doing consulting work on Apollo. "When Mueller came on board, I continued on in essentially the same job for a while," Low said, but it wasn't the same. Unlike Holmes, Mueller had very little use for him. The situation clearly had changed in Washington—and not for the better, as far as Low was concerned. "Mueller and I did not hit it off very well," he said. It was a personality and style clash of the highest order. Low operated around a team model; Mueller was a "my way or the highway" kind of guy. "He did not need or use a deputy. Mueller more than anything needed staff assistance, not a guy like me who could be his alter ego or work in certain areas in his own right. He just didn't need me."[55]

The friction between Low and George Mueller is one of the very few such contentious relationships that can be found between Low and upper management at NASA. Low credits Mueller, like most people, with pushing the all-up testing method on the Saturn V rocket, which helped the program advance against its aggressive timeline. Rather than the slow, methodical, iterative test regime that von Braun and his team in Huntsville preferred, all-up testing sought to fly all the rocket stages, systems, modules, and the like in one live test. All-up testing of the booster was, indeed, a bold and important contribution to the success of Apollo. It dramatically shaved costs and development time. There was no taking that away from Mueller. But Low is critical of both Mueller's management style and his overall technical judgment. "Mueller should get the credit for all-up testing," Low emphasized in interviews.[56] Indeed, Low believed Mueller's initiative saved tremendous time and money in the development process of the Saturn V. "It's a proper approach with the launch vehicle, since it worked," Low surmised, raising his eyebrow, when it was said to him that Mueller wanted to do all-up testing not just with the launch vehicle but also with the spacecraft itself. "I argued with him against it on the spacecraft," Low said. "A spacecraft is more than just testing a machine. It's also testing a man," and the parameters were different.[57] So, too, were the costs in the event of a catastrophic failure. Problems with both the command module and the lunar module after the *Apollo 1* fire would prove out Low's argument.

In another interview with Robert Sherrod in Washington, Low's frustration with Mueller's style flared. At one point in the conversation, he waved off a point about Mueller by saying, "By that time, I really didn't care what he thought."[58] When Sherrod mentioned that Mueller claimed to be the one responsible for bringing the independent-minded centers together, Low strongly disagreed. "It was Brainerd Holmes who brought the centers together," he insisted. He pointed out that Holmes got Gilruth and von Braun together—apparently no easy task. Holmes assigned roles and missions clearly, Low explained, which was something that Mueller did not do. He did not respect the lines of responsibility throughout the organization. Sherrod then asked Low about Gemini—someone had told Sherrod that Gemini was Mueller's "prime achievement." Low grimaced. "He clearly does not like him," Sherrod wrote in his notes. "Chuck Mathews deserves the whole credit for running Gemini," Low said. "And we had to fight Mueller on many other points— doing EVA on *Gemini 4* rather than later, for example. He almost lost that one for us." Low was the principal initiator of the EVA on that early flight.

Shortly before the first Gemini flight, the cosmonaut Alexey Leonov performed the world's first EVA, which stands for "extravehicular activity," or an activity performed outside the spacecraft—yet another first for Russia in the space race. Not wanting to be outdone yet again without a quick response, Low secretly had a team see if Ed White and Jim McDivitt could perform an EVA on their flight, the second manned flight of Gemini.[59] After a few days, flight operations confirmed that it could be done. Despite resistance from Mueller (who didn't think the astronauts could be ready in time) and Dryden (who worried that it would be viewed as just a publicity stunt), Low pushed and succeeded in getting the EVA approved for the flight.[60] Paul Haney, the head of public affairs in Houston at the time, recalled the *Gemini 4* EVA decision: "I was amazed at the breadth of the plan in Low's mind," he said. And Haney was just as impressed by Low's grasp of every detail. "Though the idea of it was only twenty-four hours old, and not yet on paper, Low outlined the testing that was needed and massive amounts of other detail," Haney said.[61] After the mission, Low was among the first to watch the developed film from the mission—and he was thrilled by the amazing images of Ed White floating outside his capsule, the blue ocean water of Earth miles below him.[62] It is some of the most stunning and popular footage from all the Gemini missions, and it proved to the world that the United States was rapidly starting to catch up with the Soviets in capability.

"Mueller was divisive," Low explained to Sherrod, trying to sum up his appraisal of him as diplomatically as possible.[63] The Gemini team, under Chuck Mathews, was working through the chain of command in Houston, pulling together all the center's resources. Because of that style, it was possible to make such a quick—and successful—decision on the *Gemini 4* EVA. Apollo, on the other hand, was operating outside the center's chain of command and competing with the center. The approach was anathema to Low. "Mueller was strong on technology, and he was inventive, but he was weak on management and future planning."

NASA future planning also concerned Low. Given the constantly changing fortunes in Washington, he knew that it was important for the agency to have detailed plans prepared in advance in order to seize moments of opportunity with the White House or Congress. Low believed the ideas being generated in the Advanced Applications Program (AAP) office under Mueller were, at best, far-fetched; at worst, they could never be funded or realized. At the time, AAP wanted a fully reusable shuttle, with projected development costs of $4 to 6 billion; an Earth-orbit space station, with development costs estimated at $6 billion; as well as a lunar space station and base, a manned Mars mission, and a growing list of manned and unmanned scientific programs. Taken together, almost every major project was a budget buster. At its peak during this period, NASA's annual budget touched $5.9 billion in 1966, when Apollo was at the pinnacle of its build-out and represented the majority of NASA's budgetary dollars. By 1970, just six months after the landing of *Apollo 11*, the budget was already scheduled to be slashed by 37 percent to $3.7 billion, and it was scheduled to continue to fall.[64] Look at Mueller's AAP plans in that light; Low viewed all of them as nonstarters as they were being discussed. "My concern was purely budget," Low explained.

> Remember, I was in a job where I spent days worrying about the 1964 and 1965 integrated budgets we had and, before flying to the moon on *Apollo 11*, deciding whether we had to rebuild the whole LM because some of that very old-fashioned equipment was suddenly coming up with a new disease that was not working. . . . And I could see a George Low in 1975 sitting in a future shuttle meeting, saying, "But it won't fly, and you can't fly it! All this wonderful stuff isn't good if you can't pay for it." I just said, "You guys are just talking through your hats. You will never build anything like this."[65]

Low was not one to gossip behind anyone's back or talk badly about them. If he didn't agree, he had no problem looking someone in the eye and saying it. Mueller was no exception. "For the record, I received a call from George Mueller, who was most disturbed about the feedback he got," Low wrote in a memo to Gilruth. "Specifically, I was accused of having stated that AAP could not be done because it was too complex, its schedules were too ambitious, and the costs were either too high or too low. I again told [Mueller] what our position had been and indicated that we urged full support to early, simple AAP missions, with emphasis on the Earth resources program."[66] Low's insight would prove correct, again in the post-Apollo period for NASA, when extreme budget cuts and a general lack of political and public support demanded more for less out of NASA and more Earth-focused benefit from its activities.

From Low's perspective, the final selection of the LOR mode made in the Brainard Holmes management committee meetings was one of the crucial technical decisions in Apollo. And it taught him to avoid going with Mueller's smartest-person-in-the-room decision-making process. "An overriding factor in the success of Apollo was the fact that the method selected for the lunar landing was the lunar-orbit-rendezvous mode," Low points out. "John Houbolt's analysis of that mode, together with his pressure in calling it to the attention of NASA decisions makers, was in my opinion a most important factor in our selection."[67] Low was initially against the LOR option. He had asked Max Faget to review it as part of the Low committee deliberations. Faget had decided, and Low's committee had agreed, that LOR was not the proper choice. "None of us were smart enough to question him hard enough on it. It was a mistake on my part not to probe further and to turn Max around and to get Houbolt directly to present to the committee." As a result, Low's committee ultimately focused on the technically complex and more expensive direct approach mode. "It is my opinion to this day that had the lunar-orbit-rendezvous mode not been chosen, Apollo would not have succeeded. . . . Other modes would have been so complex technically that there would have been major setbacks in the program, and it probably would have failed along the way." Low's reliance on a single committee member's analysis of a crucial decision, instead of opening it up to the broader committee to review, study, and debate, was an important lesson. It informed his desire to always get as many inputs as possible on any key technical decision.

Shortly after Mueller's arrival in Washington, Gilruth offered Low the number two spot, as his deputy director. Jim Elms, Gilruth's current right-hand man, was leaving for a job in private industry at Raytheon. With Shea coordinating Apollo via Mueller in Washington, Gilruth needed someone to interface on his behalf with Chuck Mathews and help run the Gemini program.[68] Since it was the program currently flying, it needed the most attention. He also needed Low to serve in a role like he did during Mercury as the STG representative in Washington—being his right-hand man, working with people, balancing the budgets and managing the money matters, and running the Manned Spacecraft Center (MSC) day to day.

"I jumped at the offer," said Low. "I think it was sometime in 1963. It became official in early 1964. I actually held both jobs for a while. In fact, I was writing letters back and forth to myself. I was officially Mueller's deputy in Washington and Gilruth's deputy in Houston at the same time. The staffs on both sides got a real charge out of writing letters back and forth for my signature."[69] When the role in Houston became official, Low and his family moved down to Houston in 1964, three years after Kennedy's dramatic go-for-the-moon decision. He was so thrilled to finally be involved in the field that he and Mary R. planned to live there permanently. Houston was to become home.

On paper, the move seemed to some to be a demotion; some even called Low, asking, "What happened?" But for him it was a liberation. "I had fun at MSC," Low recalled with fondness. "I learned a lot about running an organization. Remember, the most I had ever handled in terms of people before was a branch at the NACA—thirty people working for me directly. And I enjoyed that as much as the technical aspects of running an organization. I enjoyed working with people—making them produce more than they thought they could."[70] At MSC his responsibilities would be broad and varied: "I was responsible for a large number of things. The various programs, their budgets, and the emphasis those budgets reflected, the man power, the future, and the total resources available for the various projects."[71]

Being at a field center, he could also step out of his purely managerial roll and get involved with the hands-on, day-to-day activities. He was an early riser.[72] After his ritualistic morning run, he would meet with Mathews for a half hour on Gemini, before the day even started for the rest of the center. Sometimes he would provide advice, but mostly, he would just listen and absorb the details. He met less frequently with Shea on Apollo, who tended to work more directly with Mueller back in Washington. He'd attend Gemini flight press conferences—sometimes with Mary R. along as well. He'd show up in the middle of the night in mission control during flights to monitor progress, ask his usual questions, and just keep a watchful eye on things. And when necessary, he'd get involved with flight operations when he really felt strongly about an issue, such as pushing for the first American EVA on *Gemini 4*.

When Low moved down to Houston, it was during a time of massive growth, as the center expanded in both size and operational capability to realize Apollo's goal by the end of the decade while still finishing out the Mercury and Gemini programs. When it first opened, it consisted of the original STG team of 750 members; by the end of the Gemini program, the center had grown to almost fourteen thousand civil servants, contractors, and NASA personnel. MSC was the central hub for the design, development, and operations of Mercury, Gemini, and Apollo. As deputy director, Low was once again back in the heart of the action. And he was once again working for a man he respected and admired. "Bob built the best organization in NASA by bringing the best people into it," Low said. He surrounded himself with engineering and management talent. While Gilruth was a great technical leader, he was not a great hands-on manager. For that, he needed Low. "Being what Bob described as his general manager was a very, very outstanding and interesting experience. I hope I brought a little bit to the organization, and I learned a great deal."[73] By this time, Low was a central figure in the golden age of spaceflight, doing his part to help the nation rapidly advance toward the ultimate goal of landing a man on the moon. At MSC it was once again like the early days at the STG—a group of young engineers pushing at the boundaries of their skills and capabilities, working together as a team on something that meant more to them than just about anything else. From the first Mercury launch, through to the last launch of Gemini, NASA had begun to successfully rack up one success after another—not only approaching the Soviet Union in space capability and flight records but also, by mid-Gemini, surpassing it. Their success was fueling their optimism. Getting to the moon by the end of the decade seemed more and more like a foregone conclusion. Until, of course, the tragic *Apollo 1* fire.

4

The Longest Days

On the evening of 27 January 1967, Low was working late (as usual) in his office on the seventh floor of MSC.[1] He was plowing through end-of-the-day notes and memos while monitoring, via a direct audio feed, the *Apollo 1* spacecraft testing being conducted at the Cape. It was supposed to be a simple, fifteen-minute, plugs-out test, which meant that the crew was inside, the hatch was sealed to sixteen pounds per square inch as it would be on the launchpad, the capsule was operating on its own internal power supply for the first time, and the test was being conducted across the entire Apollo communications system—from the command module to launch control at the Cape to mission control in Houston. It was to be an almost-identical test to the one the backup crew had performed successfully the day before but with two key exceptions: it wasn't running on an external power source via cables that snaked out of the capsule through the open hatch, and it was pressurized to sixteen pounds per square inch of pure oxygen like it would be on launch day. Unlike the prior day's test, this one was experiencing a series of communication and technical problems and delays, and the "simple" test was stretching out longer than anticipated. In fact, the crew had been sealed within the capsule at 2:42 p.m. eastern time that afternoon, and it had now sailed well past dinnertime.

Joe Shea, the Apollo program manager, had just arrived in Houston from the Cape. He was scheduled to be inside the capsule for the test, laying under the astronaut couches at the bottom of the capsule, but engineers had not been able to get the communication systems working to link him with the rest of the crew and mission control. Gus Grissom, the mission's commander, wanted Shea in the capsule so he could experience firsthand the many problems they were having with the machine, the least of which was the poor communications system. This particular capsule, Spacecraft 012, had arrived at the Cape with over one hundred "significant" engineering change orders that had not yet been completed.[2] The capsule was a mess. From Grissom's perspective, the spacecraft was nowhere near flight worthy, and time was running out to fix it for its scheduled February flight. The fact that Shea could not participate in the test was simply the tip of the iceberg on problems with the command module, which spoke volumes about its other problems.

Clearly frustrated, Shea flew back to Houston to monitor developments, from his office. Not long after, at just a few seconds past 6:30 p.m. eastern time, both Shea and Low heard the first, grim inklings of the tragedy that was about to unfold: someone yelled, "Fire!" After a few frantic moments and phone calls, both Low and Shea rushed across the lawn of MSC toward mission control. Cognizant of their role as management, they took their usual place in the back row behind the flight director, as the mission controllers frantically assessed the dire situation. The unthinkable had just happened. All three *Apollo 1* prime crew astronauts—Gus Grissom, Ed White, and Roger Chaffee—were dead. They had burned to death in a flash fire fueled to over one thousand degrees Fahrenheit within the highly pressurized, pure oxygen environment, most likely started by a simple spark underneath their couches— right in the spot where Shea was supposed to be as an observer, where he had been sitting before his communication loop failed.

It was a grisly, tragic reminder for the space program of the dangers and the risks. NASA had so far flown sixteen manned flights, for a total of over 1,023 hours of flight time, under the same conditions, with no loss of life. They had succeeded in making spaceflight seem routine. But this was different. This happened on the ground. The rocket hadn't even been ignited. "We always expected that we'd lose at least one mission before we landed on the moon, because of how far we were reaching out," Walt Cunningham, *Apollo 1* backup astronaut, said, echoing the feelings of those in the Astronaut Office at the time. "But we didn't expect it to be on the ground."[3]

"I was appalled," Low said, underscoring the series of mistakes that led up to the fire—with none more critical than not managing and controlling the multitude of engineering change requests.[4] The fact that *Apollo 1* went into the test with over one hundred critical change requests still not having been made was unacceptable. "We made a serious mistake, and it took the lives of three of the finest men I have known."[5]

The fire brought about a screeching halt to the rapid pace of development and expansion in the space program. It cost three precious lives, incurred additional cost overruns of more than $135 million, and set the program back more than a year and a half. It also marked a pivotal change in the safety culture and the go attitude within the agency to beat the Russians to the moon at any cost. "It was not only three astronauts who died on that evening in January 1967," Charles Murray and Catherine Bly Cox write in *Apollo*. "Some of the space program's lightheartedness and exuberance died, too. Things had been changing all along... but the fire was a demarcation of the loss of innocence. For many of the Apollo people, the night of the fire was also when they stopped believing in their invulnerability."[6]

The ultimate causes of the fire, as it would be reported by the NASA accident investigation committee, were not limited to a single mistake or cause. There were just too many, and it involved more than just engineering. It involved poor processes, lack of management control over technical change request, as well as pressure to meet launch schedules and delivery deadlines. Low explained,

> The fire happened not because of any one thing. It happened because of a number of things that went wrong. First of all, it shouldn't have happened. It could have happened in Mercury or Gemini, but it didn't. Looking back, we did not do enough work on flammability in a sixteen-pounds-per-square-inch oxygen atmosphere; there's no question about that. It's all of us at MSC who should have done something about it. I think we were extraordinarily unlucky that the fire started when it did. We had run many tests in that environment previously, and we'd never seen a spark or an arc or anything else.[7]

Organizational complexity played a role as well. The recently completed Mercury and Gemini programs were managed with a relatively small number of people. It was easier for management to have oversight and stay on top of critical issues and program changes. "Once we got into Apollo, we suddenly had a national mobilization, and it was big," said Glynn Lunney, original STG member and NASA engineer and flight director during Apollo. Clearly, it was bigger than the current structure was able to support. "Everything was big. And there were contractors for everything, and too many people. We went from small teams to thousands. There was a lot of overlap."[8] As problems surfaced, as they inevitably do in developing new spacecraft, engineering decisions were being made that had unintended and unstudied impacts; other technical decisions were also being deferred, especially if looking into them might impede or impact delivery schedules. As a result, these issues—both large and small and seemingly ever growing in number—compounded, snowballed, and created disaster.

Immediately after the fire, raw emotion overpowered logic, and anger and blame festered quickly in Washington and the press. It also propagated throughout NASA and the contractor supply chain. Engineers started blaming each other or themselves. To compound matters, the press and politicians in Washington began clamoring not only for a cause but for people to blame and shoulder responsibility. Shea became a common scapegoat in some circles. He bore the brunt of the criticism, and he carried perhaps the heaviest emotional burdens for failing to safely manage the program.[9] He also suffered from a classic case of survivor's guilt, having escaped death because of the faulty communication system. As a result, he fell into a deep depression. "A lot of us had a sense of guilt. I've had a pretty big monkey on my back for a while," he would later say.[10]

Low saw it differently. Everyone working on Apollo contributed to the fire; there wasn't just one person to blame.[11] Wally Schirra, who would command NASA's return-to-flight mission on *Apollo 7* in 1968, concurred. The high-speed program, which put them on a crash course to the moon before the Soviets, blinded them to the problems that were festering out of control before their eyes. Success in Mercury and Gemini also dulled, it seemed, awareness for hazards—especially of the risk of a fire in a pure oxygen environment. "I've always considered the accident a total failure of us all," Schirra said, echoing Low's assessment.[12]

If Apollo was to continue, a fix needed to be found. That would have to start with management. Low and Gilruth thought Mathews would be the ideal candidate to step in for Shea. After all, he had just successfully navigated the challenges of the Gemini program, and he had the full faith and trust of both men. But Mathews had already moved to Washington for a new job under Mueller, and he was loath to uproot his family once again and so soon. In turning the job down, he let Low privately know that another name was being floated around in Washington by Webb, Mueller, and Seamans for the job. It was the only man they felt could get the job done. Their man? Low. Stunned, he and Mathews discussed the role and its importance.[13]

Some might have rejected such a job as a step down, going further away from the chain of command—a career in reverse, so to speak, after having moved from arguably the top post in manned spaceflight in Washington to a deputy center director position and now down again to that of a project manager. But not Low.

The fire was different for everybody, depending on where you stood. People in Washington—Administrator Webb and others—looked at it from the point of view of saving the agency. From my point of view in Houston, it was the most challenging time in my life as an engineer. I was assigned the job of redesigning and rebuilding the spacecraft. I would work around the clock, with a team of wonderful people, both in government and at the contractor plants, to find out what else lurked in that spacecraft, what else could come to bite us. The president had said, "Let's land on the moon before the end of the decade." And that was a challenge that meant a great deal to all of us. We didn't want to let the country down. Time was running out. From April 1967, when I took over the project, to the end of the decade, we knew the moon was only going to rise and set thirty-three more times in that period. It was a very countable number of times that the moon was going to be in the right position. The challenge of doing that, the challenge of the hands-on engineering job, the challenge of working with the team that was involved in all of that was, to me, the most fascinating job an engineer could ever have.[14]

He was officially offered the job in a dramatic and cinematic fashion. He was sitting aboard *NASA 2*, a Grumman Gulfstream jet, with Gilruth.[15] They were on the tarmac at Washington National Airport, waiting for takeoff clearance to return to Houston. Seconds before takeoff, the pilot got notification from the tower: return to the terminal and have the passengers wait in the pilot's lounge. Not long afterward, a car came screaming up to the airport, and four NASA officials got out and rushed to the lounge. "Soon arrived administrator Jim Webb; his deputy Seamans; Mueller, the head of manned spaceflight; and Apollo program director Sam Philips," Low said, describing the scene. "Counting Bob, everybody in the NASA hierarchy between me and the president was there." Webb cut to the chase. Apollo was faltering. He listed everything from the fire to the national deadline to land on the moon. NASA needed Low for the job. "They all told me that it was absolutely essential for me to take this post for the sake of Apollo," he said, and he was pressed for an immediate answer. "I probably would have liked more time to think about it, but since everyone I might have wanted to consult was already there in the room, there was no point in waiting."[16]

After his appointment, the mood and the attitude around Houston changed. For Low, it would mark the beginning of the hardest but most satisfying two years in his life as an engineer. The new job would propel his already time-consuming work habits into overdrive. "Somebody once figured out my average working day was like sixteen to eighteen hours per day, and maybe another twelve on Saturday, and more on Sunday afternoon," Low said.[17] When the raft of memos and updates got to be too much, instead of slowing down, he decided to speed up—he took a speed-reading course to up his pace. "When he became Apollo program manager, he would come to work at seven o'clock in the morning to compensate for the different time zones so he could talk to the Grumman people in the eastern time zone when they got in at 8:00 a.m.," explained Jerry Bostick, who worked with Low. "And he would stay late past 7:00 p.m. so he could talk to the people at Rockwell on the West Coast at the end of their day."[18]

It was a work schedule that he leaned into; it charged him as much as it taxed him. It was also daunting by any measure. "Two machines, 17 tons of aluminum, steel, copper, titanium and synthetic materials; 33 tons of propellant; 4 million parts, 40 miles of wire; 100,000 drawings, 26 subsystems, 678 switches, 410 circuit breakers," Low said, summing up the overall makeup of the command and service module (CSM) and the lunar module (LM), which, together, composed the Apollo spacecraft. As the new program manager, he would have to "leap upon this fast-moving train, learn all about it, decide what was good enough, and what wasn't, what to accept, and what to change," he recalled. "In the meanwhile, the clock ticked away, bringing the end of the decade ever closer."[19]

"Selecting George Low as the program manager was a stroke of genius," Lunney declared. He famously once described Low as a great "knitter of people," someone who would bring order and harmony out of the chaos and doubt caused by the fire. "George brought the program out of despair and brought it back into the light."[20] He was "a master at getting people to work together, creatively channeling their energies and thus building the momentum to achieve the objective," legendary flight director Gene Kranz said in describing Low's impact on the program. "The flight directors knew him well from his middle-of-the-night visits to Mission Control during a flight, where he sat silently in the viewing room. He had a rare blend of integrity, competence, and humility. You would do whatever he asked you to do, regardless of the odds and regardless of the risks."[21] Astronaut Frank Borman agrees with both Lunney and Kranz: "As usual with any great endeavor, it always boils down to a single human being who makes a difference. In the case of Apollo, the person in my mind who made the difference was George Low."[22]

He was a leader, in the classic sense. People followed him. Julian Scheer, the head of NASA Public Affairs in Washington, received a call from a friend who was working in Houston after Low was made Apollo program manager. This person spoke of Low's ability not only to lead but to also inspire the engineers under his command: "I hope you don't think I'm nuts," the caller told Scheer, "but I want to tell you something. We had an Apollo briefing in the ninth-floor conference room. Low spoke to us. God! It was the most exciting, lucid, thrilling, dynamic thing I've ever heard. Honest to God, it was just fantastic. He didn't even have a note. He held us all spellbound. . . . So help me, if he had said, 'Let's go,' I'd have followed him right off a cliff!"[23] And follow him they did. "Gave a pep-talk to all ASPO personnel and subsystem managers in the auditorium," Low wrote Gilruth on his second day on the job, in his typical understated fashion. "My friends tell me it was quite well received."[24]

"George was a stabilizing force in everything we did, after the fire," Gerry Griffin said. "He was a special talent. He would hear people out. He had this ability to get everybody's input, synthesize it, and make a quick decision. He was the right guy to lead us. He had a calming effect on everyone."[25] Kraft believed Low to be the ordained man for the job, by virtue of not only his innate talents—"He was a brilliant engineer"[26]—but also his prior experience: "I can't emphasize enough how extremely important he was to the situation, and from every point of view: from the technical point of view, from the management point of view, and from the political point of view, because he had experience in all of those areas."[27] *Time* magazine put it more succinctly: "George Low brought a cool discipline to the nerve-racking operations in Houston."[28]

One of the first things Low sought to do was to integrate the Apollo program office with the activities of MSC. He had to get control over the highly complex and widely dispersed Apollo supply chain and dysfunctional decision-making. It would require hands-on management that would be inclusive, rather than exclusive, as it was during Shea's tenure. According to Low, the problem with Shea's management of the program wasn't one of technical ability or of any type of overt malfeasance; rather, it was one of poor communication and organizational structure at the highest levels. Still, he refused to outright blame Shea. "George was protective of Joe," Abbey explained. "He put most of the blame on George Mueller being the cause of the communication problems, as Mueller made the arrangement with Joe to have a direct communication link with him instead of going through Gilruth."[29] For example, Mathews communicated with Low daily about Gemini, and Low kept Gilruth informed, which allowed both to bring in the resources of the entire center to help address issues. Shea tended to get inputs from the contractors, outside consultants, the astronauts, and the center leaders; however, he often took the opinion of outsiders over that of the center staff and rarely, if ever, closed the loop on his decisions. "There were things that Joe tried to do which resulted in head-on opposition from the astronauts or Chris Kraft," Low pointed out. "He was unable to convince them he was right and could order them to do things a certain way only so long, since they are people who want to know *why* they are doing what they are doing; unless they understand, the organization and the work suffer. It certainly didn't help MSC or the Apollo program to have a program office which felt its job was to pick and choose from among several conflicting opinions coming to them from inside and outside NASA."[30]

In hindsight, Low admits, he could have and should have done something about it at the time, but he did not. "Joe and I were friends, and I had full trust in Joe," he said.[31] Even when others would complain about Shea to him—such as Kraft, Faget, or Slayton, all of whom brought very specific Apollo issues and concerns to his attention—he always referred them back to Shea. Shea was the man in charge of that program, and Low respected the chain of command. Rather than micromanage Shea, he would tell them to try to work it out with him first. "From a personal point of view, I liked Joe, too," Kraft said, trying to explain why this approach just didn't work. "But I was having problems with him. He would not respond to our inputs. He would ask around for our opinion but not wait for the answer. He would just make his decision. If he didn't like something, he would go down to the contractors and have them do what he wanted them to do. In effect, he was pitting parts of the organization against each other."[32] After much effort, Kraft believed that he was starting to make inroads and that Shea was starting to come around just before the fire. "I spent a lot of time trying to convince Joe that he was ignoring our inputs and that he needed to take advantage of the tremendous engineering experience and talent that we had at MSC. I think he had begun to realize that he needed a lot of help and that he could get it from us. I think our relationship was changing."[33] Unfortunately for the crew of *Apollo 1*, it was too late.

Even though Low was not involved in the day-to-day engineering of Apollo, he allowed his trust in Shea's technical knowledge and capabilities to blind him to the fact that basic project management mistakes were being made. "What I learned is that you've got to dig in a little more deeply as management and look over people's shoulders, even if you think they are the best in the world at what they do," Low said. "You've got to penetrate a little more deeply. My way of doing business at the time was not the right way for understanding what Joe was doing. Bob and I let people have their way, and we didn't penetrate, because we expected our people to come to us. Joe didn't bring us in. He did not cut us out, and we didn't try to penetrate. We didn't insist on being cut in. We should have."[34]

By penetrating, Low was referring to one of his core tenants in engineering and program management, tempered into an unbreakable rule after the fire: making sure that people are asking the right questions in order to get to the right answers. "We had our test procedures before the fire," Low said as a matter of example. "We just didn't ask ourselves all the right questions."[35] Case in point, the spacecraft hatch. During the fire, the astronauts could not egress from the spacecraft fast enough after the fire ignited. The hatch, as designed, needed several minutes to open—too long for the oxygen-soaked environment that caused even the aluminum tubing to burst into flame, let alone all the netting and combustible material within the capsule. After the fire, the hatch was redesigned. It would open outward, instead of inward, and it could be opened in an emergency in a matter of just five seconds, which he described as now being able to be opened "with your little finger" with the help of a cylinder of compressed nitrogen gas.[36] "Before the fire, we had designed a hatch which was a perfectly good engineering design, especially if you are given the problem of designing a hatch that's absolutely safe in space: you design one that opens inward, it's lightweight, it won't fall out in space, and it will seal against the atmosphere."[37] This was the model for the hatch in the *Apollo 1* spacecraft. Yet if you ask yourself the right question, he posited, such as, "Can the crew quickly and safely egress the spacecraft on the launchpad if there is a fire during a test?" you would find that they couldn't, of course. They were locked into the spacecraft. It wasn't a matter of poor mechanical engineering; it was a matter of not incorporating human-factors design *into* the engineering.

In the process of breaking down the next Apollo spacecraft scheduled to fly—and now asking the right questions in light of the tragedy—Low and his team discovered, and ultimately fixed, more than 1,407 different errors and defects in the one spacecraft.[38] "When I became program manager, the next spacecraft—CSM 017—was in terrible shape," Low said. "I sat down with Joe Bobik, who was chief inspector at the Cape. He had been a crew chief at Lewis in Cleveland, where I had known him. Joe was the kind of on-the-ground-guy I would sit down with before every flight and I would ask, 'Joe, what do you think of it?' I did that throughout Apollo. I asked Joe, 'How could spacecraft 17 happen?' And he said, 'You got what you wanted.'"[39] It was a simple yet damning answer. NASA wanted a quick job—by the end of the decade— and they also wanted quick deliveries at low cost. This meant less testing and more tolerance of errors, without properly thinking through the problems they might cause. It also caused people like Shea and Mueller to cut communication corners and seek outside help over the center, and it caused people like Gilruth and Low to accept it that way.

"How do you get what you want in this business?" Low asked, rhetorically, after relating the Bobik story. "You get what you want when a Deke Slayton (or a Chris Kraft) comes to you and gives you an opportunity to complain, and you don't accept his complaint. Instead, you say, 'Go see Joe Shea.' That's how you get what you want."

Not any longer. After the fire, Low's rallying cry amounted to asking questions and penetrating until he got a satisfying answer. "We asked questions, received answers, and asked more questions," he said. "We woke up in the middle of the night, remembering questions we should have asked, and jotted them down so we could ask them in the morning. If we made a mistake, it was not because of any lack of candor between NASA and contractor, or between engineer and astronaut; it was only because we weren't smart enough to ask all the right questions."[40] After the fire, Low insisted that every question be answered, every failure be understood, and every problem—no matter how big or how small—be solved.

"Before the fire, the Apollo program was insulated," Borman explains. "It was not integrated enough into the center, and a lot of the decisions were made in Washington. That just doesn't work." Empowering local questioning and decision-making was key to getting the program back on track. The information should come from the ground up. Borman, who was a fan of Low's management style, would go on to emulate it as CEO of Eastern Airlines. The approach was further solidified for Borman when he consulted with one of von Braun's colleagues after the fire. "I remember a long conversation with Dr. Alexander Lippisch," Borman said about his time working with Low and others to fix the Apollo spacecraft. Lippisch was a legendary German aeronautical engineer brought over with von Braun during Operation Paperclip after World War II to White Sands Missile Range. Inspired to go into the field of aviation after seeing a Wright brothers flight demonstration, Lippisch is the engineer who designed and built the first aircraft to fly under rocket power, and he designed the Messerschmitt Me 163 rocket-powered plane. "He told me that you have to make certain you don't run into what happened in Nazi Germany, where all the decisions were made in Berlin. 'Remember this, Major Borman,' he told me in his German accent, 'Make the decisions at the lowest level where the information exists. Don't become focused on the Berlin complex.' If you look at the Apollo resurgence after the fire, and Apollo's ultimate success, it was primarily due to the people at the center—people like George Low—who made this happen."[41]

To further integrate center leadership into the process, Low brought Gilruth back into the chain of command. "MSC is Bob Gilruth's center. He built it in terms of what he felt was needed to run a manned spaceflight program," Low said. "Bob has ideas. He inspires confidence. He knows what is right and what is wrong. He also expects the rest of us to originate ideas and carry them through to completion."[42] As Apollo program manager, Gilruth was Low's boss, and not Mueller in Washington. Low needed to have Gilruth as part of the team, and he needed to keep Gilruth fully informed and to be his buffer with Mueller. "Because of our schedule, it was impossible for me to talk daily with Bob," Low said. "But I did write, for every day I was on Apollo, a daily note that was usually about two pages long which I dictated first thing in the morning when I got to the office. I told him on a daily basis what was going on in Apollo. Nobody else saw those because I wanted to be able to tell him about people, about contracts, and I wanted to feel perfectly free. It was a private discussion with Bob."[43] This daily interaction was an important part of his process. "In every major decision I make, I ask myself, 'Is this the way Bob would make the decision?'" Low said. "I am close enough to him and keep him well enough informed on what we do, day-by-day, that he would have every opportunity to disagree with me if he chose to do so. He seldom does, because I bounce enough ideas off him." Writing his daily memos also helped him organize his thoughts. "When I first came in to the office in the morning, I took ten minutes organizing what happened to me before and what I was going to do that day before I dictated those notes."[44]

He also believed strongly that the center director needed to be more involved than just being an informational conduit to Mueller. He needed to *own* pieces of the program. "I felt that I not only needed Bob's support but that he ought to have part of the program. He ought to feel responsible for it," Low said. So he put Gilruth in charge of all the acceptance reviews of the spacecraft when it would arrive from the contractors, as well as the flight readiness review process. Shea had reserved those reviews for himself. "I asked Bob to chair those reviews," Low explained. "It was my way of seeing that Bob and the center became involved in those things, that he had to take responsibility for acceptance of the spacecraft from the contractors."[45]

Despite reestablishing Gilruth into the chain of command, George Mueller tried to go around him and go directly to Low. It was part of Mueller's divisive management style. Besides not being his boss, Mueller was also not Low's only main contact with Washington. That interaction was formally with Sam Phillips, who reported to Mueller. "Sam and I had an outstanding relationship," Low said.[46] "I would talk to Sam on the phone, probably every day. He and I knew that I would keep him informed and he could always get into the act down in Houston if he wanted to. So that was the kind of manager he was. He trusted you. He delegated, and he trusted us to keep him informed. The more things bothered me, the more I would tell him about it and get him to question the decisions. And that was a very good and open information relationship." Early on in Low's tenure as program manager, Mueller called Low. "He went into some detail and asked me something. I said, 'George, I will find out for you.' Then I hung up, went to see Gilruth, and I said, 'Call back your boss, George Mueller, and tell him about it. I don't have a channel to George Mueller.' I cut it off that day, and it never opened up again."

But that didn't stop friction from occurring between the two throughout Low's service as Apollo program manager. Mueller would criticize Low's additional tests and work, second guess his approach to engineering on the spacecraft, and try to backdoor him with the contractors. In his notes, Low often underscored the impact these clashes had on confusion with the contractors, as well as on production and flight schedules. "Lee James called and informed me that he spent most of the morning discussing schedules with George Mueller," Low wrote to Gilruth. "He therefore will not accept my schedules."[47] In another note, he writes that "George Mueller is most upset about the strong position that MSC is taking on CSM delivery schedules. Part of the problem is that George thinks that North American is telling him that they can do better than MSC says they can do. I have a complete matrix of all schedule positions and will be prepared to discuss this in Washington if necessary."[48] It is clear from Low's notes that he never let the pressure from Mueller's office in Washington divert him from what he thought was best in terms of safety, schedule, or chain of command. He certainly wasn't afraid to hop on a plane to Washington to set the record straight if Gilruth wasn't able to contain Mueller.

The establishment of Low's configuration control board (or CCB, as it was known) was the next major milestone in changes made at MSC. The CCB served as a central mechanism to bring all MSC and contractor resources to bear on fixing the myriad Apollo problems. "Apollo was successful because of the configuration control board Low put into place and how he managed it," Kraft agreed. "It is the only way we were able to get anything done. He wouldn't tolerate any nonsense out of the contractors, either. He was a hard-nosed guy when it came to that. It made all the difference in the world."[49]

"By April of 1967, when I was given the Apollo spacecraft job, an investigation board had completed most of its work," Low said. "The board was not able to pinpoint the exact cause of the fire, but this only made matters worse because it meant that there were probably flaws in several areas of the spacecraft. These included the cabin environment on the launch pad, the amount of combustible material in the spacecraft, and perhaps most important, the control of changes. Three mistakes, and perhaps more, added up to a spark, fuel for a fire, and an environment to make the fire explosive in its nature. And three fine men died."[50]

The CCB would change that dynamic.

Low came upon the CCB solution when the concept was suggested to him by Shea's former assistant, George Abbey. Abbey, a thirty-five-year-old engineer and a former U.S. Air Force pilot, began working for Shea in 1964. During his time as Apollo program manager, Low would keep Abbey on as his technical assistant, and he developed a trust in Abbey very much like he had for people like Kraft or Slayton. "The CCB was organized by my technical assistant, George Abbey, who knew everything about everybody on Apollo, and who was always able to get things done," Low said.[51] He leveraged Abbey's immense Apollo knowledge and skill similarly to how Silverstein and Gilruth had employed Low during Mercury and Gemini. As Low's eyes and ears on the ground, Abbey would be able to do the all critical administrative groundwork that Low himself could not get done.

"The two Georges were a remarkable team," said astronaut T. K. Mattingly. "George Abbey had this network of working people. He knew the troops in all the buildings, and he'd wander around and just talk to people and bring all of that stuff back. He would just make sure that Low was aware of everything going on as perceived from the bottom of the barrel." In Mattingly's mind, this access to the informal MSC grapevine helped prepare Low in the asking of his famous impromptu questions in management meetings, because Abbey was able to bring to his attention matters in a diplomatic fashion that normally might not filter up to the top. "George Low is the finest program manager that ever walked the face of the earth," Mattingly said. "I've seen a lot of good people, but I've never seen anybody of his caliber. He had this way that he could take information and use it but never embarrass anybody or never expose. He would just ask questions, and the source of a lot of this came from George Abbey."[52]

"We were talking in his office one morning," Abbey said, recalling Low's formation of the CCB. "'I've got to overcome the problem Joe created when he set Apollo up as a kind of separate entity within MSC,' he said to me. 'I've got to somehow meld the Apollo program office into the center.' I suggested that he set up a configuration control board and make the directors of each of the major organizations be on the board, as well as the program manager from each of the contractors."[53] Low liked the idea, especially in light of the success he had experienced as part of Brainard Holmes's management committee meetings. He viewed the board as not just a place to discuss and adjudicate the appropriateness of changes; it also served a much broader purpose. It provided a decision-making forum for the three legs of the Apollo stool: spacecraft manager, spacecraft builder, and spacecraft user. "In reaching our decisions, we had the combined inputs of the hardware developer, flight operations, flight crew, safety, medicine, and science," Low said. It was a clearinghouse for bringing both NASA and contractor management together; still, it did not relieve Low of the burden of being in charge. Each decision was made in front of the full CCB by Low, not by the group. "Decisions were not made by the CCB, as the decisions have to be mine," Low said. "I make them after hearing everybody's opinion for or against, but I do not take a vote on it. I make the decision."[54]

"George got us all into a room, and he said, 'I'm putting this board together,'" Kraft said. "All of you guys—the leaders. You will *all* attend. No substitutes. I don't want anyone but you. You need to be on board 100 percent, because, together, we are going to run this program."[55] The CCB team included the following: his two deputies, Ken Kleinknecht and Rip Bolender; Apollo's assistant for flight safety, Scott Simpkinson; Max Faget, Houston's chief engineer; Chris Kraft, the chief of flight operations; Deke Slayton, the head of the astronauts; Dale Myers, for North American Rockwell (manufacturer of the CSM); and Joe Gavin for Grumman (manufacturer for the LM). Low considered all these people to be some of the best engineers in the world; he respected them and their opinions.

To round out the engineering, he also included Dr. Charles A. Berry, director of medical research and operations, and Wilmot Hess, director of MSC science. The contributions of the science and medical teams were sometimes as important as that of the engineering. Having this team together helped solve, for example, the flammability problem within the 100 percent oxygen environment of the spacecraft, which had been used in all successful U.S. spaceflights until the fire. "We solved the problem of fire in the space atmosphere of [five pounds per square inch] of oxygen," Low said.[56] "But try as we might, we could not make the ship fireproof in the launch-pad atmosphere of [sixteen-pounds-per-square-inch] oxygen. Then Max Faget came up with an idea: launch with an atmosphere that was 60 percent oxygen, and 40 percent nitrogen, and then slowly convert to pure oxygen after orbit had been reached and the pressure was [five pounds per square inch]." The CCB needed the expertise of the medical and science members because "the 60-40 mixture was a delicate balance between medical requirements on the one hand (too much oxygen would have caused the bends as the pressure decreased), and flammability problems on the other." Together, the medical and engineering teams figured out the right mix, and the atmosphere solution worked throughout the rest of the program.

In order for the CCB meetings to be effective, Low knew that clear agendas had to be set, a chain of command had to be established, and people with decision-making authority needed to be at the table. Without that structure to a meeting, there is no accountability, and nothing can get done. He had experienced this early on in Apollo during one of the first North American management review meetings in September 1962. Charles Frick, who was the first Apollo spacecraft program office manager, invited Low to Downey, California, for a North American management review meeting. Low was less than impressed at the time.

"It is my understanding that the purpose for the monthly meeting is to give you, as Project Manager, the type of information you need in order to assess the status of the work," Low wrote to Frick. "My general feeling of last week's meeting is one of disappointment. I feel that neither your people nor the North American people gave you the type of information that you really need to stay on top of the job."[57] Low was also concerned that no one was in charge of the meeting and that a chairman had not been established. For good measure, he quietly asked those in attendance who was in charge; no one was able to tell him. He was also not impressed with the caliber and type of information presented. "I believe the status review . . . should be one of the most important features of the meeting. . . . North American should be asked to compare recent progress against predicted curves. This information should be presented in enough detail so that you can obtain a clear indication of impending slippages or cost overruns at the earliest possible time. I think the information presented . . . was unsatisfactory for this purpose."

Since Low was not Frick's manager, he did not press the issue but, rather, simply gave the information to Frick as a courtesy. "I would like to emphasize that I am sending you this letter as a personal note," Low wrote Frick. "Not as an official letter. If you disagree with any of the comments I have made, I will not feel at all disturbed if you disregard them." Of course, this kind of optionality on Low's observations was no longer valid after the fire. The CCB, and the decisions Low made in those meetings, ruled.

Low charged Abbey with writing the CCB agendas for the meeting each week, which would take place religiously every Friday and start at exactly noon. The meetings often went well into the night. "We dealt with changes large and small, discussed them in every technical detail, and reviewed their cost and schedule impact," Low said. "Was the change really necessary? What were its effects on other parts of the machine, on computer programs, on the astronauts, and on the ground tracking systems? Was it worth the cost, how long would it take, and how much would it weigh?"[58] The agenda would get posted on a Monday, and all the parties involved had the next four days to prepare for the meeting. "Those people in charge of an area we were discussing would have to come address that problem on Friday with a solution; if they didn't have a solution, they had to bring a plan on how they were going to get a solution," Abbey explained.[59] In the early days, the hallways were filled with grumbling that people did not have enough time to prepare; as the weeks turned into months, the grumbling took on a new tone—people were now pushing to have their items added, not excluded, from the agenda.

The length of the CCB meetings was breaking one of Low's normal protocols of having no meeting last longer than forty-five minutes. The issues at hand, how they were explored in-depth, and the sheer volume of the change requests, all dictated that the meetings go until every issue of the week had been hammered out. Given the emotions sometimes involved and the broad discussions, Abbey said that Low used a gavel to call the meetings to order, and he would sometimes use the gavel to gain order if an argument broke out. People were allowed to discuss issues—even passionately—but he would not tolerate any chaos or disrespect. At the end of each meeting, Abbey would write up the meeting minutes for Low's signature and distribution. "And those minutes—including the decisions that he made in the meeting—would go out as a directive right away to everyone in the program," Abbey said.[60] Once a decision was made, after everyone said their peace, everyone rallied behind the decision and did not second-guess it or try to subvert it. Low would not tolerate any subversion of CCB decisions. "George would let everyone have their say. Not everyone agreed on everything. We had diverse opinions. Some of the participants got very emotional about what they wanted or how they saw everything. George would listen to all of it. He gave them the opportunity to say what they had to say. And then he would make his decision, 'This is what we are going to do.' And we did it."

If anyone disagreed with Low's decision, they were invited to take it up with Gilruth. According to Low, no one ever did, mainly because the decisions were made in the open and with full transparency. "Arguments sometimes got pretty hot as technical alternatives were explored," Low said. "In the end, I would decide, usually on the spot, always explaining my decision openly and in front of those who liked it the least. To me, this was the truest test of a decision—to look straight into the eyes of the person most affected by it, knowing full well that months later on the morning of a flight, I would look into the eyes of the men whose lives depended on that decision. One could not make any mistakes."[61]

Low of course knew his own limitations. He was not a man to think that he knew everything, and despite getting the best advice and inputs he could find, he sometimes would second-guess himself or have his doubts. He was always cognizant of the dangers of the smartest-person-in-the-room syndrome. When he had doubts, he turned to Gilruth as not only his boss but also his longtime mentor and friend. "When I wasn't sure of myself or when I didn't trust my judgement, I knew where to go to get help. Bob had been through every problem. He had acquired great wisdom over the years dealing with men and their flying machines. He was always there when I needed him."

The agendas for the CCB meetings, the minutes of the meetings, and the decisions were all published and circulated. They were posted on a billboard in the lobby for all to see—including visitors and the press. He even invited journalist and NASA historian Robert Sherrod to be an observer at meetings. With Low, radical transparency and full information sharing was a key to getting everyone rowing in the same direction. It was also critical in the oversight of every piece and part change involved in the redesign, retest, and rebuild of the spacecraft. Low later wrote,

There is an old saying that airplanes and spacecraft won't fly until the paper equals their weight. There was a time when two men named Orville and Wilbur Wright could, unaided, design and build an entire airplane, and even make its engine. But those days are long gone. When machinery gets as complex as the Apollo spacecraft, no single person can keep all of the details in his head. Paper, therefore, becomes of paramount importance: paper to record the exact configuration; paper to list every nut and bolt and tube and wire; paper to record the precise size, shape, constitution, history, and pedigree of every piece and every part. The paper tells where it was made, who made it, which batch of raw materials was used, how it was tested, and how it performed. Paper becomes particularly important when a change is made, and changes must be made whenever design, engineering, and development proceed simultaneously as they did in Apollo.[67]

In addition to diving deep into the paperwork, he believed there was no substitute for on-site inspections. A couple of weeks into his tenure as program manager, he began scheduling weekly site visits to the two main contractor sites—flying out to Downey, California, to meet with the people at North American, and out to Bethpage, New York, to meet with the people at Grumman. He would also frequently fly to the Cape for CSM and LM checkouts, as well as shuttle back and forth to Washington with Gilruth for senior NASA management meetings and updates.

At North American, Low was appalled at what he saw on his first visit—from poor workmanship and attitude to understaffed and unqualified workers being assigned to the job. After the visit, he wrote to Gilruth, "My general impression after this week's visit is that Dale Myers, Charlie Feltz and George Jeffs are trying extremely hard to do the right things. Their attitudes are proper, and they are working hard to deliver us a good Command and Service Module. The next level below them, however, disturbs me."[63] Low was particularly concerned with the team in charge of spacecraft wiring. Faulty wiring was a principal cause of the outbreak of the fire on *Apollo 1*, and in the past week, Low's team had identified over one hundred faulty switches that were "found to have a failure mode which could be quite catastrophic." All the switches had to be replaced. Low wanted an update on wiring at this particular site visit, but he was less than impressed with the team leader from North American. "Unfortunately, he had a rather flippant attitude about the job," he wrote. Another manager struck him as "just plain no good." He noted that the environmental control system engineer was "certainly not on top of his job" and that the manager in charge of spacecraft weight "certainly didn't know the spacecraft weights, nor did he appear to care. . . . From all of this it looks like my next effort will be to try to get better people."

To ensure that his directives would be correctly implemented, Low deployed key NASA people on tiger teams at the various contractors to oversee the work being performed and to report back to him. He often paired astronauts with MSC engineers. In the case of North American, the contractor building the CSM for Apollo, Low put Frank Borman in charge of the on-site NASA team. "He sent me and a team down to North American, and we monitored the changes that George was making to the Apollo program as we reengineered it. As the proposals came in to modify the Apollo, we kept track to make sure they were being implemented correctly. Also, I reported to George on North American management and practices, and I was his eyes and ears out at North American."[64]

Borman also reported back to Low about the rather "schizophrenic" atmosphere at North American—a mix of conscientious, dedicated employees who knew what they were doing "and at least an equal number who didn't know their butts from third base."[65] Low did not waste any time in implementing personnel corrections at North American. According to Abbey, he moved swiftly when he thought workers were either not competent in their jobs or were in over their heads.

He was not very patient with individuals who didn't know their job. Every time we addressed a problem or an issue on the spacecraft at the change board, the engineer who was personally responsible at either North American for the command module or Grumman for the lunar module would have to make the presentation. He required those individuals to make the presentation—no others, no substitutes. He wanted to be sure that those individuals that were working on that system really understood the system and could explain it to his satisfaction, and also explain why a problem existed and explain the solution. I saw him fire people on the spot, people he felt were not competent. He didn't go by the person's age, or maturity, or who they were—he strictly went by a judgement on whether or not they understood and could explain the system they were working on.[66]

Of course, the Apollo spacecraft included not only the CSM but also the LM. When Low took over, his assessment of the LM being built by Grumman was almost as bad as it was with the Rockwell-built CSM. "In LM-1 more than 1,000 squawks [problems] have been found in the wiring inspection alone," Low noted. "Many of these are substantive, including bare wiring terminals, broken insulation, and the like."[67] Low would go out to visit Grumman on 12 April 1967, and he would report that "the scheduling and understanding of work that needs to be done are pretty much out of control." He noted that "significant technical problems include a very tight weight situation" as well as an increase in the number of LM-1 problems to 1,200 "crabs." "Crabs," Low wrote, "is the East Coast version of squawks."[68] Even the vocabulary between the main contractors for the same issue was different. They were all speaking a different language.

Despite the increase in "crabs," Low remained perhaps a bit more optimistic at this early stage of the LM than he was with the CSM "squawks" out at Rockwell. "LM-1 probably looks much like Gemini and does *not* look like CSM 017," Low wrote. Still, he noted serious concerns on production scheduling, which he deemed "the weakest point" and in which he had "no confidence whatsoever." As time progressed and he spent more time with Grumman and the LM, his doubts grew. "As in the CSM, the *initial* picture is not very encouraging. Much has been done over the past several years that I just cannot agree with," he wrote.[69] Issues included things like redundant wires passing through a single connector—meaning that if the connector fails, so, too, would both the main and the redundant backup wire. The problems would only continue to grow from there. Low wrote Gilruth as late as 29 November 1967:

> Technical progress at Grumman is still slow. I keep having the feeling of being in a debating society meeting as opposed to dealing with a technical group of people. Vibration testing at the component level is finally getting defined after many months of "leading by the hand." I made several speeches, ranging from the need for Company management to back its Vehicle Managers on the schedules they are proposing to a desire to see more initiative on Grumman's part, as opposed to NASA

leading Grumman by the hand all the time. I pointed out that a good manned flight organization would be coming to us with solutions of all technical problems, as opposed to trying to rationalize them away.[70]

"The LM was the first true spaceship," Low later wrote. It would be tucked into a cocoon within the Saturn V during launch, because it could only operate in space. Unlike the CSM, which had to ballistically navigate the earth's atmosphere to land, the LM could only operate in the vacuum of space. "Designed by aeronautic engineers who for once did not have to worry about airflow and streamlining, the LM looked like a spider, a gargantuan, other-world insect that stood 23 feet tall and weighed 16 tons," he said. "Like a spider, it was somewhat flimsy, with paper-thin walls and spindly legs."[71] It was like nothing any engineer had ever built before; as such, problems were to be expected along the iterative steps necessary to make the machine flight ready. But the problems at Grumman were exasperating for Low; these problems would ultimately lead to the long delay in LM deliveries.

As with his many trips out to Downey to inspect the CSM, his weekly trips out to Grumman in Bethpage, New York, indicated the need to have a NASA tiger team in place there too. He selected Jim McDivitt—an aeronautical engineer and Apollo astronaut whose technical expertise, management ability, and flying skills he trusted—to be his astronaut eyes and ears on the ground with the tiger team in Bethpage. He saw in McDivitt a person who was hard driving, as well as someone who not only understood the engineering but also had a broad perspective on the entire program. It is no coincidence that McDivitt also commanded *Apollo 9*, the first test flight of the LM in earth orbit, and it is also no coincidence that Low picked McDivitt to replace him as Apollo program manager after *Apollo 11*. Low pushed management at Grumman, as he did at Rockwell, and he insisted on improvements to the schedules. When there were problems on NASA's side about which the contractors complained, he gave those equal weight and insisted on maximum effort from both sides. But he would not allow anything on either side to override flight safety or a CCB directive.

To further engage the contractors at the highest level, Low himself rotated the CCB meetings once a month between Grumman and Rockwell. He was also known to show up at various times during any given week, unannounced, just to check up on things. Not just management by looking around, but management by surprise. In a Johnson Space Center *Roundup* piece, Low was remembered during this time "as the relentless top NASA official who was often seen at the contractor facilities on weekends."[72] One time, Low showed up at 8:00 p.m. in Downey and spent two hours walking around at night to see how work was being performed after hours. It was this kind of attention to detail, he would point out later, that often eluded so many other senior managers who chose to stay behind a desk and not want to get their hands dirty. For him, there was no substitute for seeing the shop floor in person. During this particular late-night visit, he stumbled on to some of the reason for the slowness in getting the spacecraft out of the factory. "This was most discouraging in that very little activity was evident. I saw two people in the Command Module of 101, two in 2TV-1, and two in Spacecraft 020. There should be five or six in each. A lot of people are standing around outside the spacecraft apparently waiting for equipment, tools, brackets, or something," he noted. "On the stack of Spacecraft 020, several people were looking for two brackets which were taken out of the spacecraft some time ago and could not be found now when they were needed for reinstallation." After the site visit, he could make specific, highly effective recommendations to Rockwell management on how to improve work-floor performance. His observations got attention and immediate action: "Apparently, we really got North American's attention, last week, after I visited the plant during the night shift," Low wrote to Gilruth. "I have letters from both Meyers and Bergen on this subject. I am also told that Atwood raised hell after my visit."[73]

In his own calm, confident, and in-charge manner, he began to whip the program into shape. He knew the value of simply being on the ground—of being seen—and asking questions, of pulling the team together, rather than pitting one person against another. Eventually, Low's approach rubbed off. It became the culture. "Participated in subcontractor motivation program held at North American," Low wrote to Gilruth in June 1967. "Reviewed the spacecraft schedules with Myers, Feltz, and others. A tremendous amount of progress has been made since my meeting last week. . . . Charlie Feltz has now moved into building 290, with an office overlooking the checkout area. . . . This means Charlie and his top people spend a lot more time on the floor than they had ever done before. I took a brief walk around the checkout floor myself and was quite impressed with what had been done with the Spacecraft 101 wiring."[74] Low was not just the Apollo program manager. He was the eyes and ears of the workers, and he became the agency's conscience and role model.

"I did pride myself in giving as much of my job away as I could, though," Low explained, referring to his willingness to delegate important tasks to those he trusted and respected and to those who he knew from experience over the years could get a job done.[75] He could delegate this way because he had put the organizational and management structures in place via the CCB and other actions to overcome the issues experienced under Shea's leadership. Low institutionalized not only collaboration but also a system of checks and balances. Key people in Low's orbit at the time whom he trusted implicitly included people like Abbey, Borman, Faget, Kraft, Slayton, and a host of others.

"Frank Borman, who was later to command *Apollo 8*, was my alter ego in the Rockwell plant down in California," Low said, underscoring his trust in Borman. "He and I were on the telephone daily, and I visited out there weekly. Together, we worked ourselves out of the problems. I think we jointly came up with what had to be done, the new materials, the new hatch, the endless hours of deciding what kind of an atmosphere we should launch with, and how could we launch with it. How can we get the flammable materials out, how do you cover things so they won't burn."[76]

"For example, one of the things that burned profusely in the Apollo fire was an insulation that we decided to put on every spacecraft after Gordon Cooper's *Mercury 9* flight," Low explained. "In that flight, droplets of water had collected in the weightless condition of space, behind the instrument panel, and it shorted out some of the instruments. We then specified that we had to insulate everything better. Unfortunately, that insulation turned out to be flammable in the Apollo's 100 percent oxygen atmosphere. We had to reinvent a new insulation. Through testing, we found that the insulation we then invented absorbed water. So, we had to reinvent again. It was a joint effort between us and the contractor to find something that didn't burn."[77]

During this time period, a *Time* magazine writer described Low as being "reserved and distant" and "not a humorous man."[78] Nothing could have been further from the truth. Most people who didn't know Low did not know that he had a wonderful, dry sense of humor. This humor would often manifest itself at key moments to break the ice, or just when the mood struck him. He would write funny notes and observations to others during meetings. This humor comes out occasionally in Low's notes to Gilruth. After a particularly long note detailing problems that arose while Gilruth was out on vacation, Low ended the note with "Welcome back!" He also reserved his humor for those he was particularly close to, such as Borman. "Because Borman's head is considerably wider than most, we have to procure a special helmet for him," Low wrote to Gilruth. "This is still the same kind of bubble helmet we are using, but it will be an inch or two wider around the ears. When you see Frank, you might inform him that his big head is costing us $45,000."[79] His family saw his humor more often than most. A favorite memory from his children was from a time when he tried to teach them how to taste and order wine at a restaurant. At their own dinner table, he dramatically went through the motions of opening the bottle, airing out the wine, sampling a sip, and then proclaiming it worth drinking. "What if you don't like it?" he was asked. Going through the motions again, he crinkled his face and simply spit the wine on the floor.[80]

Serious? Yes. Humorless? Not in the least.

Another classic case of Low's delegation of responsibilities involves project management of the computer program software that had been within the Apollo office. The software in the CSM was code-named Colossus, and the software in the LM was called Sundance. Both Colossus and Sundance were slow in coming out of MIT, where it was being written and coded, mostly because of all the change requests that were coming in to them from NASA, the contractors, and the astronauts. His notes to Gilruth in July 1967 increased in urgency about his software concerns. "It appears that once again we are developing real scheduling problems at MIT, and many of the programs are slated to be available much later than desired. This problem will be compounded if we make a significant change in the use of the spacecraft computer as suggested by Tom Stafford. (Stafford has recently completed a study which indicates that much simplification is necessary in order to avoid too many pilot inputs to the computer.)"[81] With constant user change requests, MIT could not stabilize the scope long enough to finish the programing. Without the software, Apollo could not fly. By September 1967 MIT's schedules had slipped another six months behind.

Low took Kraft with him to MIT to assess the problem in early October.

It was worse than either had thought. "It was terrible," he confirmed. "Visited MIT with Kraft. . . . We had hoped that by today MIT would have determined a schedule; would have prepared a configuration management scheme; and would have a viable organization. Unfortunately, however, it apparently takes so long to get anything moving that very little had been done." Letting MIT self-manage the problem was not going to work. They were too deferential to the astronauts' every desire. He needed someone to get in there and straighten out the mess. "That night, in Jim Elm's daughter's bedroom—we were staying the night there at Jim's home to save money, and since Jim's daughter was away at college—I turned to Chris and said, how would you like to take over the software on Apollo?"[82] Low felt that he could not handle yet another task on his plate as big as the software component, in addition to the physical hardware on the spacecraft. Given Kraft's skills in mission operations, as well as his understanding of the astronauts' needs, Kraft was a natural for it. Plus, he had Low's complete faith and trust.

"I am not a brilliant software guy," Kraft said. "I didn't particularly want that job. But I said okay, and I took it. And I didn't question it, because it was important for Apollo and it was what George needed." It didn't take Kraft long to spot the problem at MIT.

> Every time MIT tested, they had so many changes, so many astronaut wish list changes. MIT couldn't get it done, because it couldn't get finished. With the software, you had to damn well make sure that any change didn't impact or change anything else. And better is always the evil of the good in that situation. The question to ask, which no one until I got there had the authority to ask, is, Does the software work without the change? Not, Does the change make the software do something "better"? Stop trying to make it better, or you'll never finish. Just make it work. The MIT guys worshiped the astronauts, and they wanted to give them their every wish. It was a domino effect, but it ground everything to a halt. The change requests were the problem. So I set up my own CCB, like George had in Houston. I gave MIT thirty days to solidify the software in current state, to figure out their existing issues on hand. After thirty, I told them, there would be no changes without my signature. The MIT guys appreciated it. It put in a process, and it created a buffer and a control between them and the astronauts. And

from then on, the software started to come out of there like sheets of paper on a printer.[83]

Low was pleased. He knew Kraft could do it, even if Kraft himself had doubts. "I completely handed Apollo's software over to Chris," Low proudly said. He later listed the fixing of the software as one of the top things necessary to get Apollo back on its feet. "And I never bothered with it again, because once a week Chris had his people report to me on overall progress of their work. I gave it to a guy that I knew could do it better than I could."[84]

In the end, Low's Apollo spacecraft CCB met ninety times between June 1967 and July 1969. During this time, 1,647 changes were considered for both the CSM and the LM, and 1,341 were approved.[85] "The low rejection rate resulted because proposed changes were reviewed before they came to the board, and only those that were deemed to be mandatory—for flight safety reasons, rather than to simply make things 'better'—were brought before the CCB for discussion," Low explained. "We tore the command module apart—literally all 2 million parts—and then we put it all back together the way we wanted to."[86]

The redesign did not come without its own cost, in terms of dollars as well as weight. In fact, the total weight of the spacecraft increased by 3,500 pounds, which pushed Apollo to a weight that was only 350 pounds under the Saturn V's total lifting capability for a lunar mission—even with von Braun's extra center engine.[87] This heavier weight meant that Low needed to redesign Apollo's parachute system to be able to handle the increased weight load and that some redundant systems had to be eliminated. Metals were switched out—such as the spacecraft's heavy lead ballast—which helped eliminate four hundred pounds. Even some metal brackets inside Apollo had holes drilled into them to save additional ounces. All of the changes meant an increase in both quantity and frequency of testing each and every time a new change was ordered.

"Our underlying design philosophy was to use redundancy wherever possible, and to provide the simplest possible interconnections among various systems," Low said. The goal was to reduce, rather than increase, complexity. In fact, minimizing the functional interfaces between the spacecraft and the Saturn V launch vehicle brought the number of interconnecting wires down from over one thousand to just one hundred. "A single engineer can fully understand this interface, and he can cope with all the effects of a change on either side of the interface. If there had been ten times as many wires"—such as the fifteen miles of wires just within the CSM—"it probably would have taken a hundred times as many people to handle the interface." Together, redundancy and simplicity "made for a very forgiving design: many things could go wrong and often did without endangering the mission or the safety of the crew. We recognized that components would fail. Statistically, there were too many of them for this not to happen. We designed the system so that a component failure could be tolerated."[88]

"Following the fire, we made changes, looked at the spacecraft, examined it, tested it, and we found other things wrong, too," Low said. Building something as complex as the Apollo spacecraft can only be done in iterative steps. Designs are made and built, but then they must be tested. Most problems, he said, were only discovered once a system was placed in use. "We had a lot of problems with titanium tanks, so I learned a lot about titanium tanks," he said. "We had less problems with steel tanks, so I did not know as much about them."[89] The only way to truly shake down such a machine was to actually use test articles in extensive flight simulations and to put them through the same rigors they would experience during a mission. This took considerably more time and money than had originally been planned, but it was critical to Apollo's eventual success.

> We were especially good in the extra care we took in testing after the fire. Before the fire, we might have had a 90 percent system, and then had a 99 percent system afterwards; or a 99 percent system before the fire, and a 99.9 percent system after that. You could put whatever numbers you want on it. But that extra-special look, taken by so many of us, following the fire— the failure modes and effects, sneak circuits, special reviews, special drop tests, more and more testing of individual switches and components— all of those things, together, made it possible to go to the moon. I am not sure whether, without those things, we would have gotten there.[90]

In Low's mind, the most important thing about management and running Apollo was the attention to detail: "I believe in people who know how to do their jobs and are willing to do them. No detail was too small for anybody to look at. That was probably the most important management principle involved in Apollo."[91] In his daily notes, he continuously uncovered new problem areas that were not being addressed and then quickly went about setting up teams to fix them. These problems could be as deadly as the slow opening hatch or a high-oxygen atmosphere or critical components installed backward or simply in the wrong place. "Circuit-breakers remain a large open item," he wrote. "They are highly flammable and can conceivably be self-igniting. Furthermore, they are not temperature compensated and since they involve a hot-wire element, they are pressure and 'g' sensitive."[92] In other words, they were designed for use in space and not in a one-g environment on the ground—like the hatch. Faulty circuit breakers, sticky and malfunctioning valves, and poorly wired switches would generate major problems for Low and the team—so much so that he would keep a table in his office with all the problem switches and circuits they discovered.[93] He kept them there as a reminder of all that could go wrong and of all the little details that needed constant scrutiny, reinvention, and retesting. "Each represented a potential failure in flight," he said. "Many represent a crisis before a flight—a case where we did pay attention to detail and where we may have saved a mission."[94]

This focus on detail did not just pertain to the engineering; it extended across all his responsibilities. He expected everyone at MSC and across NASA who worked on Apollo to focus on the details as much as he did. Doug Ward, who worked in public affairs at MSC, had joined NASA just six months before the *Apollo 1* fire. Ward recalls that Low was very adamant that he wanted to review all press releases, press statements, and any communication as it related to Apollo. He viewed the details and accuracy of those releases to be just as important as any engineering report or study.

I had a very privileged view of Low, because one of my first assignments was to cover the work coming out of the engineering director, as their public affairs representative. Low made it clear that he wanted to see it before it went out. I'd get a release pretty well finished and then march up to his office. And his secretary had instructions that when I showed up—whatever he was doing—she would get me in to see him. And he would interrupt important meetings with all kinds of high-level people and come out into the outer part of his office to review my news releases. He always did everything in green ink, and I had all of these news releases that were replete with his little additions and corrections. And I never questioned him, because it was so obvious from the nature of his changes that they were not only constructive, but they were right.

He was a genius at perceiving public perceptions and was particularly adept at knowing what kinds of information needed to be developed.[95]

Low wasn't just concerned with every word in NASA press releases. As an avid photographer, he was also deeply interested in the photos the astronauts returned from their missions. He felt they were a powerful part of the scientific and public benefit being returned by the space program. Throughout his life, his home office was often adorned with large-format prints of iconic mission photos—such as Ed White's space walk on *Gemini 4* or *Apollo 8*'s *Earthrise*. The images were signed by the crew, and they were dedicated to him. Many are on display in the George M. Low Gallery on the RPI campus. One of his favorite images was an iconic photograph of the LM floating in Earth orbit during *Apollo 9*. It was the LM's first manned test flight, and its success ensured the rapid move toward the lunar landing on *Apollo 11*. Jim McDivitt, *Apollo 9*'s commander, inscribed the photo to Low: "Many thanks for the funny looking spacecraft. It sure flies better than it looks."

"Low played a big part in selecting the photographs that we released," Abbey confirmed. He appreciated the power of the images to inform and sway public opinion. "As soon as we got film back to Houston, George wanted me to call as the films were developed. He would arrive right away to look at the pictures, every time. He would go through the photos. And if he saw one in particular that was really an outstanding photograph, he would say, 'This is one we certainly need to get out to the public.'"[96]

He was also a central figure in pushing for getting whole-Earth photos from as early as possible in the Apollo missions. He knew the public would be enthralled to be able to see the whole planet from space for the first time. "We are making efforts to get a camera into Spacecraft 017 so that we can take Earth pictures from the 9,000 mi. apogee," Low wrote to Gilruth. "We should be able to see the entire globe with about 25 percent in sunlight and the rest in darkness. This had not been planned until this time."[97] His efforts resulted in the beautiful *Apollo 4* crescent-Earth photo taken during the unmanned *Apollo 4* test flight in November 1967. The photo was taken when the *Apollo 4* spacecraft was orbiting Earth at an altitude of 9,544 miles. It was taken with a 70mm camera that was programmed to look out a window toward Earth, snapping a picture that showed swirling white clouds and deep-blue oceans on a quarter crescent of the earth. NASA officially issued the photo to the public as image number AS04-01-580. In later manned missions, whole-Earth photos would become even more dramatic—helping, many argue, to inspire the environmental movement of the 1970s. On *Apollo 17*, the last Apollo lunar landing mission, the *Blue Marble* photo was taken, becoming one of the most iconic images of the earth from space.

As a passionate advocate of human-factors design and of the critical role the astronaut performed in flying the spacecraft, Low also kept a keen eye on details of the astronaut's flight-operation procedures. "George was very concerned about flight procedures and the checklist the crew used, and the checklists corresponding with how the system operators worked in the control center," Abbey said. "There was concern that the checklist used by the crew and the checklist used by the people in the control center were not really maintained and kept up-to-date and were different." Low was most concerned with the activities in flight where the crew would use a procedure in the spacecraft that was different than the way it was understood in the control center. "The people in the control center were very concerned about making sure that the crew's procedures were in sync with their procedures. Once George was aware of their concern, he got concerned. He had me go down to the Cape with a team of people two days before every launch and go through the crews' checklists step-by-step. Sure enough, we found differences."[98]

Of course, some of Low's attention to detail required him to go the extra mile himself. Once, he tried to convince the astronauts to use a small emergency oxygen supply of just fifteen minutes during the return from the lunar surface to the CSM in orbit. He wanted the astronauts to abandon a much larger, heavier breathing apparatus on the moon. This would give them more space and weight capacity in the LM for lunar samples and get them more mileage out of their remaining fuel. The astronauts resisted—and not for inconsequential reasons. What if the LM failed to dock with the CSM? they argued. What if they had to do an EVA from the LM to the CSM in an emergency? Fifteen minutes would not be enough time. "No way," they told him. A few months later, he returned to the discussion with the astronauts. He brought along a film to show them a scuba diver in the MSC EVA training tank conducting the maneuver using the fifteen-minute oxygen supply. The diver made the transfer between the LM and CSM mock-ups effortlessly underwater and with plenty of time to spare. Impressed by the skill of the diver, the astronauts still balked. They pointed out that it was perhaps easy enough for a professional scuba diver to do such a maneuver underwater, but the astronauts were not professional scuba divers. The diver in the film? That was no professional diver. "That was me," he told them. He had taken a crash scuba course just to make his point. If he could do it—the dirty-hands engineer who had never dived before and wasn't an astronaut, just a desk jockey—then it could easily be accomplished by them. Impressed, the astronauts agreed. And they flew with the fifteen-minute tank.[99]

Stepping in and going through the same simulations as the astronauts was something that Low, in particular, enjoyed doing. From the human-factors approach, this not only helped him understand their perspective; it was also a hell of a lot of fun. At times, he comes across in his daily notes like a young teenager whose father is letting him drive the family car for the first time: "Went to the MIT Instrumentation Lab for a lesson in celestial navigation, using the Apollo guidance and navigation system," he wrote. "MIT has a complete LM guidance system set up in a radar gimbal mount on the roof of one of their buildings. I did several platform alignments by making star sightings with the telescope and sextant and using the on-board computer. My first alignment was accomplished with a zero-error indication by the computer. All of this made me feel pretty good, especially since McDivitt was there and saw it all."[100]

In another pertinent example, he got into the simulators to understand clearly what the astronauts were requesting.

> Spent several hours with Warren North, flying pad abort missions on the AMX, 205/101 rendezvous missions on the McDonnell-operated hybrid computer. . . . It is clear that the times available for each step in the sequence are minimal; nevertheless, I was able to accomplish each abort "safely" except one where the instructor pulled a double failure on the launch escape motor ignition system. Based on this rather limited experience, I am still not convinced that the "sky-diving" technique might not be possible. . . . Incidentally, I used 250 pounds of propellant for a case where the theoretical minimum is 70 pounds.[101]

Details, details, details . . .

Even after hundreds of design changes to the CSM, the sheer complexity of the vehicle and the millions of piece parts that had to operate on any given mission—from start to finish—meant that there would be failures. This of course didn't count the equally voluminous design changes required for the LM or the work being done in Huntsville on the Saturn V rocket by von Braun's team. In fact, with over 15 million piece parts in the overall, complete Apollo stack, even with an incredible (and some would argue impossible) 99.99 percent reliability rate, there would still be about 1,500 individual failures per flight. Through rigorous testing and retesting, Low's goal was to make sure that those failures occurred in only inconsequential and redundant areas—especially in the 4 million parts that he oversaw in the overall combined Apollo CSM-LM spacecraft.

Low's meticulous but necessary rework of the spacecraft set the program back a number of months in its race against the decade's-end goal, as well as added millions to the budget. Some criticized him for the delays and additional costs. He would hear none of it. The basic probability statistics demanded no less. "I felt the things that came out of the woodwork after the fire should have really been handled before," he said. "I am not going to apologize ever to anybody about the changes we made."[102] His notes to Gilruth are replete with examples. While some changes had to do with safety, others involved basic flight operations that would have made a lunar landing impossible if they had not been addressed. Early on, for example, he notes that "communications between two people and the ground, and telemetry from both astronauts from the lunar surface is now impossible. Apparently, this is a problem that has been left open for the last two years with Kraft, Berry and Slayton making repeated inputs to the program office, which the program office has ignored."[103] Having to adjust timetables goes on all the time in any complicated technical project like Apollo, he explained, but never to the exclusion of safety. "There are three overriding principles that have to concern the project manager," Low explained. "Technical performance, i.e., will it work, and will it work safely and well. That has to come first. Schedule and cost, of course, matter, too. They come together to form the legs of a stool. But safety has to come first. If it doesn't fly safely, you won't get there at all, either."[104]

"The attitude in the Astronaut Office was that we were going to lose one of the Apollo crews somewhere, somehow after the fire," Borman admitted, summarizing the reality of the situation and the stoicism that experimental test pilots exhibited with their jobs. It is also perhaps the highest praise and testament to the results of the focus and work Low put in as Apollo program manager. "But we didn't. We came close on *Apollo 13*, but the performance of the people and the machines was unbelievable. To be able to operate a brand-new vehicle every two months—without loss of life—it was very impressive."[105] On *Apollo 7*, the first manned flight after the *Apollo 1* fire, the CSMs suffered only twenty-two relatively minor problems—or "anomalies," in NASA parlance. By the time *Apollo 11* rolled around, the CSM experienced only nine.[106] This incredibly low failure rate speaks to the herculean task that Low and his team undertook after the *Apollo 1* fire to get the program back on track and to the moon by the end of the decade.

Spaceflight—as anyone within NASA will tell you and which George Low knew as well as anybody—is inherently dangerous. You cannot eliminate 100 percent of all risks or prevent all failures. Managing a human spaceflight program is about balancing risks. It was management's job to sweat every single small detail and to work practically around the clock to make sure that critical issues and decisions were being investigated, tested, and resolved. It was attention to excruciatingly minute detail and a work ethic second to none that differentiated Low's efforts from the one that came before him. "I do not think he slept very much during this period," said Gerry Griffin, a young flight controller during the time of the *Apollo 1* fire who would go on to work closely with Low in Washington in the years preceding the approval of the space shuttle program. "He was just working all the time."[107]

In the end, he converted a faltering, fire-prone death trap into "a compact, solid, and sturdy" spaceship that could survive the searing heat of reentry—heat that could "melt and vaporize all the material in the command module several times over"—as well as travel at speeds of up to twenty-five thousand miles per hour while keeping three humans alive in space for a six-day lunar journey.[108] In so doing, he gave the astronauts—and himself—the faith that each mission could and most likely would conclude successfully. "What George Low did was instill a sense of dedication and purpose among those working under and with him," Alan Shepard recalled. "In the astronaut corps, we marveled at the new Apollo spacecraft taking shape. We were gaining confidence all the while that, yes, they're creating something that will be safe for us to fly."[109]

Between 10 April 1967, which was Low's first day on the job as Apollo program manager, and the first manned flight of the Apollo CSM during the successful *Apollo 7* mission on 11 October 1968, eighteen long, pressure-filled, and hardworking months had elapsed. As the recovery from the fire continued, progress was being made, but doubts that the end-of-the-decade goal could be met were starting to surface because of delays in hardware production—especially with the LM, which did not have its first unmanned test flight until *Apollo 5* on 22 January 1968—and the need to keep redesigning and retesting new piece parts. Progress with the redesign and testing of the CSM was going much better than expected. But issues with the LM and the Saturn V were causing concern. "The probability of landing on the Moon before 1970 is not high," Gilruth wrote in a September 1967 memo as the year was closing out.[110] Problems with the Saturn V also started to raise concerns. Despite the elation around a successful test flight of the unmanned rocket on *Apollo 4* in August 1967, a follow-up test flight on *Apollo 6* in August 1968 created a number of concerns around pogo oscillation that occurred. This oscillation is the result of vibrations in a liquid-propellant rocket engine caused by combustion instability. There were also other engine reignition issues, and several engines failed to light. "The flight clearly left a lot to be desired," von Braun glumly wrote about the mission. "With three engines out, we just cannot go to the Moon."[111]

People were resigning themselves to the idea that the goal might very well be missed.

So when George Low came back from a short vacation in July 1968 and suggested that if *Apollo 7*—the first manned flight of the CSM on the Saturn V stack—was a success, *Apollo 8* should be converted from an Earth-orbit mission into a slingshot-around-the-moon mission, the shock of his fellow engineers could be understood. Had George spent too much time in the sun? Was he serious? Could it be done?

In short order, they found out the answer.

He was very serious.

It could be done. And not only just a slingshot around the moon but an actual orbit sixty miles above the surface as well.

***

"We had a need to meet our schedule and to get to the moon by the end of 1969, before the end of the decade," Low said. "To me, that meant that we had to get a major flight off before the end of 1968." By a major flight, he meant one with enough complexity and operational payoff that it represented a giant leap forward in the goal of a lunar landing. It had to inspire and not just tick a box off a list. "As it was, *Apollo 8* was planned to be that flight, a manned flight in Earth orbit, of both the CSM and the lunar module," he said. It was to be the next step along the way, proving that the hardware could operate in space. "Unfortunately, the lunar module wasn't ready. It had its growing pains. When it got to the Cape, it would not check out as it should. It had what we call 'first-ship problems.' It always takes the first ship longer to get through, and we knew we would work out the problems, but it was just dragging along."[112]

Originally, *Apollo 8* was to have the mission profile of what eventually turned out to be *Apollo 9*—the first manned test flight of both the CSM and the LM in Earth orbit. While the CSMs for *Apollo 7* and *Apollo 8* were progressing out of Rockwell, the LM for *Apollo 8* was not. "CSM's 101 and 103 are proceeding well," Low wrote. "LM-3 is continuing to have difficulties. . . . We decided to go into the mission simulation over the weekend to see how many other problems we can identify before we go into another troubleshooting session."[113]

Besides the software problems, the LM was overweight and had a series of other production issues that kept it behind schedule. Ironically, it was the eventual commander of the *Apollo 9* mission, Jim McDivitt, who urged caution and a slower production schedule on the LM. "I spent time with McDivitt at his request," Low wrote to Gilruth in July, shortly before suggesting the *Apollo 8* moon flight. "Jim is somewhat concerned that we are pushing a little too hard on schedule. He had no specific complaints and was quick to point out that nothing has yet been done that might degrade safety. I told him that I would not do anything drastic as a result of this discussion, but that I would certainly consider his words in our daily actions. I also asked him to be sure to let all of us know if he feels at any time that we are making the wrong kind of decisions or doing anything foolishly."[114]

"The idea for *Apollo 8* was germinated in August of 1968," Low recalled. "*Apollo 8* flew in December of that year."[115] The seed of the idea to convert *Apollo 8* into a lunar mission without the LM had actually been planted in his mind in his first week as Apollo program manager, a year prior. In a meeting with Deke Slayton and Chris Kraft on 14 April 1967, the three men were reviewing CCB-approved changes from Shea's tenure. It was clear the LM schedules were slipping, and they might need to make up time lost to the fire. During the discussion, Kraft mentioned a few ideas that intrigued Low enough to write them down. One was to consider flying a rendezvous pod in the first manned flight mission—which was eventually done on *Apollo 7*, using the s-ivb stage. The other idea involved a trip to the moon: "Assuming it will take a long time to fully-qualify the lm," Low wrote in his notes, "consider a manned circumlunar mission with the csm only in our planning."[116] It was an idea that stuck in Low's head and refused to leave, moving closer and closer to the fore each time he'd get a report of more lm schedule slippages and problems. It practically screamed at him after talking to McDivitt.

Gene Kranz recalls that after Low started as program manager that April, Kraft came to a mission-planning staff meeting and had the team simulate different kinds of lunar trips around a four-thousand-mile orbit. "But it still had a csm and lm in it," Kranz said.[117] As Kranz sees it, Kraft's idea inspired Low, and this early simulation work put a lot of the data on the shelf for when Low needed a way to overcome the lm delays. This prework created the foundation that made it possible for a quick turnaround answer by Kraft and his team in August when Low boldly returned to the idea. As for Kraft, in the intervening months, he had more or less forgotten about the scenario. When Low came back from vacation in July 1968 and suggested in August a lunar mission, Kraft was absolutely stunned by the suggestion. "We were taken aback," Kraft admitted of the moment.[118]

From Low's perspective, the solution came down to asking the right question at the right time. "I asked myself the question, 'What can we do to get ourselves ahead with the hardware that is ready?'" he said. "And flying into lunar orbit was the basic idea behind *Apollo 8*. We could get a lot of problems out of the way with the flight: navigation to the moon, getting into lunar orbit, the burning of the big engine, the computer programs that were needed for that. We could get all that out of the way, while still having the time necessary to focus on solving the lunar module engineering and software problems."[119] It was the perfect technical and operational solution; it also had the potential to pack an amazing global pr punch.

On 29 September 1975 George Low wrote a memo to the nasa History Office, outlining his recollections on the *Apollo 8* decision. The office was working on a history of *Apollo 8*, and he felt they were getting some of the fine details wrong. "Although I agree with your statement that the *Apollo*

8 decision was a 'complicated business,' this should not be an excuse to 'rewrite history,'" he wrote.[120] In its write-up, the History Office implied that the reworking of the *Apollo 8* flight into a lunar mission took place only after the success of *Apollo 7*, which was factually incorrect. The reworking of *Apollo 8* had been made prior to *Apollo 7*, while the go-ahead was given only afterward.

The following represents highlights, in Low's own words, of the key events leading up to the decision to create an alternate flight plan for *Apollo 8*, making it a lunar-orbit mission:

> In June/July, 1968, the current situation in Apollo was that LM-3 had been delivered to KSC [Kennedy Space Center] somewhat later than anticipated; and CSM 103 would be delivered to KSC in late July. Check-out of 103 at KSC was proceeding well, and a launch in the Fall of 1968 appeared to be assured. There was every reason to believe that 103 would be a mature spacecraft but that for many reasons LM-3 might run into difficulties. Certification tests of LM were lagging; there were many open failures; and the number of changes and test failures at KSC was quite large . . . during the June–July time period, the project launch date had slipped from November into December, and the December date was by no means assured.

It was during this time period that Low began to think about using CSM 103 in either a lunar-orbit or a circumlunar-orbit mission, as a way "to take a major step forward in the Apollo Program."[121]

Between July and August 1968, Low wrote that CSM 103 arrived at the Cape and "the spacecraft was extremely clean." The LM was a different story. "LM-3, however, required much more work at KSC than anticipated. There was a significant number of changes in addition to test failures, requiring trouble-shooting, changeouts, and retests, and a serious EMI [electromagnetic interference] problem that continued to persist." The outlook for a launch of LM-3 "appeared to be very dim." In the coming days, they "still weren't quite on top of the situation, and the list of problems continued to grow instead of decrease." On 7 August 1968 Low formally asked Kraft to look into the feasibility of a lunar-orbit mission without a LM. Two days later Low met with Kraft to update him on CSM 103 and LM-3 and informed him that he "had been considering the possibility of a lunar orbit mission." Gilruth, Low wrote, "was most enthusiastic and indicated that this would be a major step forward in the program."[122]

Low met with Slayton the same day. The preliminary studies done back in April 1967 had indicated that such a mission was "technically feasible from the point of view of ground control and onboard computer software." Low then met with Gilruth, Kraft, and Slayton, all together. "After considerable discussion, we agreed that this mission should certainly be given serious consideration. . . . We immediately decided that it was important to get both von Braun and Phillips on board in order to obtain their endorsement and enthusiastic support."[123] Gilruth called von Braun and set up a meeting in Huntsville. Low called Phillips and Debus, and he asked them to fly down to Huntsville to join him at 2:30 p.m. in von Braun's office.

At the Huntsville meeting, Gilruth, Low, Kraft, and Slayton represented MSC; Debus and Petrone represented KSC; Philips and Hage represented the Office of Manned Spaceflight from Washington; and von Braun, Rees, James, and Richard represented Marshall. Low described for the assembled group the current situation with the spacecraft, estimating that if they waited for the LM to checkout, their mission would not fly until March 1969. "It therefore appeared that getting all of the benefits of the 'F' [lunar-orbit] mission before the 'D' [earth-orbit test flight of the CSM and LM] was both technically and programmatically advisable." Chris Kraft was the one in the meeting, according to Low, who "made the strong point that, in order to gain the F mission flight benefits, the flight would have to be into lunar orbit as opposed to circumlunar flight."[124]

After several hours of discussion, the meeting adjourned at 5:00 p.m. All in attendance "exhibited a great deal of interest and enthusiasm" for the mission that Low had described. They agreed to study it more and to meet again in Washington on 14 August. "At that time the assembled group planned to make a decision as to whether to proceed with the plans or not," Low wrote. Philips was designated to be the one to inform Mueller and Webb, both of whom were in Vienna for an international space conference, if they agreed to move forward. "It was also agreed to classify the planning stage of this activity [as] secret, but it was proposed that, as soon as the Agency adopted the plan, it should be fully disclosed to the public."[125]

The next day, Low, Gilruth, Kraft, and Slayton flew out to Washington to meet with Paine, Phillips, and a few others from NASA Headquarters; von

Braun and his team from Huntsville; and Debus and Petrone from KSC. The teams spent several hours going over the plan, with no showstoppers in the works, and suggested targeting a launch window the week of 20 December. At some point in the meeting, Low notes, George Mueller called in from Vienna.[126] Phillips had given him a summary the previous day. "After thinking it over," Low wrote, "Mueller's reception was very cool. . . . It was hard for us to believe that Mueller was unwilling to accept the plan which was unanimously accepted by all Center Directors and Program Managers." After the call, the group encouraged Phillips to continue working on Mueller to get him over his concerns.

Paine echoed Mueller's concerns. After all, he suggested, it was not that long ago that the agency was thinking of having this mission be unmanned. Now they were taking the bold step to not only man it but also take it into lunar orbit. "Have you really considered all the implications?" he wanted to know. Because of his and Mueller's concerns, he also wanted to know if anyone in the room had any objection. So they went around the table.

Low notes the following from the key people present:

VON BRAUN: Once a decision has been made to fly a man . . . it doesn't matter to the launch vehicle how far we go. . . . The mission should by all means be undertaken.

SLAYTON: This is the only chance to get to the moon before the end of 1969. It is a natural thing to do in Apollo today. There are many positive factors and no negative ones.

GILRUTH: There is always risk in manned spaceflight, but this is a path of less risk. In fact, it has minimum risk of all of our Apollo plans. If I had the key decision, I would make it in the affirmative.

KRAFT: Probably the flight operations people have the most difficult job in this. . . . It will not be easy to do, but I have every confidence in our doing it. However, it should be a lunar *orbit* mission and not a circumlunar mission.[127]

The others in the room—Hage, Debus, Petrone, Bowman, Richard, and Schneider—all agreed in the affirmative as well. Low was the last to speak. "This is really the only thing to do technically in the current state of Apollo," he assured Paine. "Assuming a successful *Apollo 7* mission, there is no other choice. The question is not whether we can afford to do it, it should be can we afford not to do it."[128]

According to Low, Paine congratulated the group "for not being prisoners of previous plans," and he indicated that he "personally felt this was the right thing for Apollo." Paine told the group he would work on Webb and Mueller. But it was clear they would also need Phillips's support. As the days progressed, Phillips "broke the log jam" with Mueller—who still had not yet talked to Webb—but Mueller wanted any public announcements and a final decision to wait until after *Apollo 7*. Planning, however, could proceed. "Phillips indicated that Webb's initial reaction was one of shock and that he was fairly negative to the proposed lunar orbit mission."[129]

After further effort from Phillips, Webb came over to Mueller's view. They could continue to plan, but they could not have a final decision on a lunar-orbit mission or make a public announcement until, and unless, *Apollo 7* was successful. "Our challenge, therefore, is to be prepared to carry out a full lunar-orbit mission without committing the agency to such a mission at this time," Low noted.[130]

"Politically, of course, it was a bad decision," Low said. "Remember, *Apollo 8* came along after the fire, at a time when only two Saturn V's had flown. The second one had several failures with it, and men had not yet flown in Apollo. So, politically, if you look at it, I'm sure that Jim Webb must have thought we all had lost our minds." In fact, Webb expressed that exact thought—rather loudly—from the conference in Vienna when he first heard the idea. In Webb's mind, the program could not politically risk another loss of life and certainly not in as high profile of a fashion as during mankind's first journey to the moon. "From my viewpoint, which is technically and maybe politically in a different sense, as in let's meet the national objective of getting to the moon before the end of the decade, it made eminent sense," Low explained.[131]

Low went back to Houston, and he settled into work on the mission planning, awaiting the flight of *Apollo 7* for the go or no go on *Apollo 8*. He discussed the proposal with Bill Bergen at Rockwell, "who appeared less receptive than most of the people who had been exposed to this plan"—some of the first cautious contractor resistance to the idea. By 13 August, planning had proceeded smoothly. "There were no 'show stoppers' in any of the spacecraft systems and, in fact, only minor changeouts would have to be made to bring the spacecraft into a position to fly the proposed mission," Low wrote. "Kraft reviewed all of the operational elements and determined that there would be no insurmountable difficulties."[132] A launch window was identified for between 20 December and 26 December. Slayton assigned Borman's crew, including Lovell and Anders, as the prime crew for the mission; Armstrong, Aldrin, and Haise were backup.

"The lunar module just wasn't cutting it, and the Cape was just driving us bananas telling us the hardware wasn't good enough," Kraft recalls, discussing why he urged Low to go for an orbital mission.[133] The men had talked about the idea of a circumlunar flight, just sending it up for a quick trip around the moon. *Aviation Week*, Kraft noted, had recently reported that the Russians would be trying for a lunar flight soon. He assumed, given Low and Gilruth's security clearances, that they had that idea in the back of their minds as well. Although the Russian consideration was part of the equation, for Low it was not the prime motivator. His big desire was to make a giant leap ahead in the schedule and keep moving forward, as well as gain important experience in the operations of actually voyaging to the moon. "It would give us a leg up on starting to improve our game plan and ability to get to the landing flight in this decade," Low told the team. "Low's idea to circumnavigate the moon was a stroke of genius," Gilruth said. "It broke the back of the Russian moon-landing effort, and it left the U.S. free to take its time and concentrate on doing the job of landing a man on the moon in the most effective way without the constraint of a deadline."[134]

Gilruth's approval was key in Low's mind.

"As soon as it was presented to [Gilruth], he said, 'Let's do it,'" Low recalled. "If Bob had told me or anyone else that he didn't think we should do it, there would have been no question in my mind, in Chris' mind, or in Deke's mind, because we would have known that Bob had a good reason for not doing it."[135]

"Low explained the mission to me and asked me to verify that the command and service modules could do that mission," Aaron Cohen said. At the time, Cohen's job was to validate the hardware for flight. Low wasn't just interested in a plain yes or no, of course—he probed with specific questions, for specific yeses or nos. Cohen remembers questions such as "Was it certified?" "Was it of the integrity that would allow us to do that part of the lunar mission?" A whole litany of questions. In the end, after his answers, Cohen ended up writing a memo to Low, certifying the vehicle for flight on the specified mission. Low gladly signed off on it. "It was a bold maneuver, a bold step from where we were," Cohen said of the moment. "I don't think I was really shocked as much as I was enthusiastic and excited."[136]

"It was the boldest decision of the space program," Kraft proclaimed.[137]

Kranz saw the decision as a clear indication that the space program had moved on from the trauma of the fire to a point of confidence in the leadership team. It was not unlike the time, he recalled, when Low pushed for the EVA on *Gemini 4*. "It was good, quality, gutsy decision-making," he said.[138]

Apollo 8 was to be an astounding flight of firsts: first manned launch of a Saturn V; first burn of an S-IVB into a lunar trajectory for lunar-orbit insertion; the first time humans left the gravitational influence of Earth; the first time NASA tried to navigate with onboard systems and software to the moon; the first time humans went into orbit around another celestial body; the first time humans looked down on the moon from a relatively short distance of just sixty miles; the first time a manned craft exited orbit from around another celestial body; and the first time a manned spacecraft would fly a thirty-six-thousand-foot-per-second reentry.[139]

With all these firsts, the mission was not without critics. Mueller believed the *Apollo 8* mission represented an "unnecessary risk" to the program. Shea was of an equal mind. Webb had left NASA shortly before the *Apollo 7* flight—in many ways because of the boldness of the *Apollo 8* mission. He just didn't have it in him anymore. He couldn't face another loss of life in the program. However, based on the preliminary meetings in August, the new administrator, Tom Paine, was all for it. And so was the rest of NASA leadership. "We had to sell *Apollo 8* to the outside world and to convince others that it should be done," Low said. "*Apollo 8* in general was very easy to sell to the other NASA centers, to Sam Philips, and to Dr. Paine. But Mueller was very much opposed to it. There have been many other decisions—the way the Apollo missions should be flown, the sequence of missions, what each step should be—and MSC has always taken the lead on these issues. As a center, it generally prevailed, more often than not against Dr. Mueller's desires."[140]

Low was intimately familiar with the spacecraft. After the studies done by Kraft, Slayton, von Braun, and others in the command chain and after all the reengineering and testing, he was convinced the mission was worth the risk. Flying in Earth orbit for eleven days or flying to the moon and coming home—heck, it was all just another trip into space. The distance itself didn't really matter, as long as the craft was flight worthy. "Granted we had analyzed, designed, tested, redesigned, and retested to minimize risk. Yet risks remained," Low said.[141]

The final decision to fly *Apollo 8* to the moon without the LM would all rest on a successful completion of the CSM manned shakeout flight test of *Apollo 7*. While test flights of the Saturn V and boilerplate CSMs had been flown—some with success and others with a series of serious technical anomalies and problems—the flight of *Apollo 7* would, itself, be a very dangerous and bold mission. Coming as it was in the shadow of the *Apollo 1* fire and commanded by the *Apollo 1* backup crew, the mission would also have to compensate for not having a LM ready for use. Going back to Kraft's recommendations to Low, it was suggested to use the S-IVB module for docking and rendezvous exercises to practice maneuvers and capability prior to having the LM available for actual in-flight testing. Despite issues with head colds and a bit of a surly commander in Wally Schirra, there were few showstopping anomalies. "Everything is going well," Low wrote to Gilruth during the flight. "I know of no *Apollo 7* anomalies that will have an effect on *Apollo 8*."[142] After eleven days, the *Apollo 7* manned mission was a successful return to flight. "We had no major problems with any of the subsystems," Abbey reported back to Low and Gilruth. "Kenny Kleinknecht talked to Wally this afternoon, and Wally indicated again that it was a magnificent flying machine and that he did not feel that there was a need to make any changes for CSM 103 [scheduled for the *Apollo 8* spacecraft]."[143]

It was time for a go or no go on *Apollo 8*.

After *Apollo 7* the final decision for *Apollo 8* would be decided in Washington at an Apollo program executives meeting with the main Apollo contractors on Sunday, 10 November, and a NASA management meeting on Monday, 11 November 1968. Early in the week, Low was concerned that the decision would be negative, especially given Mueller's and Bergen's earlier positions. "In the meantime, we are working on preparations for next Sunday's and Monday's meetings," he wrote to Gilruth. "From the number of phone calls I get from Washington, I am still concerned that things may be quite confused and not as easy as we thought they might be."[144] His worries drove him and the team toward meticulous preparation, to be able to answer and counter any negatives that might arise.

On Sunday, 10 November 1968, Phillips kicked off the contractors meeting by presenting a summary of the situation and decision to date.[145] Lee James then reported on launch vehicle status being ready for manned flight despite the vibration issues on the *Apollo 6* flight. Following the short briefing by James, Low stood up to discuss spacecraft readiness for a lunar-orbit flight. In his usual manner, he started with three questions: Is the spacecraft design adequate? Will the systems perform as designed? Are the benefits worth the risk? "I felt it was important to cast the issues in this light, since over the last several weeks we have been asked many questions that indicate that people really didn't understand that the mission we are about to fly is the design mission for the Apollo spacecraft," Low noted.[146] "It is a mission that we would have had to face sooner or later anyway, and the risk involved in performing the mission now after a successful *Apollo 7* flight is no greater than it would be a year from now." He went into meticulous detail on Apollo design redundancy and the fixing of the minor *Apollo 7* anomalies, and then he concluded by assessing the risk-reward proposition: "The risks are no greater than those that are generally inherent in a progressive flight test program. The probability of success of the ultimate lunar landing mission would be greatly enhanced."[147]

Afterward, Slayton and Kraft gave their briefings from an MSC perspective, as did Petrone on KSC readiness. Philips summarized the meeting, and then Mueller asked for a go or no-go vote.[148] Walter Burke at McDonnell Douglas was go for a circumlunar flight and not a lunar-orbit flight, but the other contractors—GE, AC Electronics, MIT, IBM, Martin Marietta, Boeing, Ford-Philco, Chrysler, United Aircraft, Grumman, and both Bill Bergen and Lee Atwood from North American—were all a solid go.

"As manufacturers of the spacecraft, our motivation to take chances is no higher than Frank Borman's, but we are ready to go," Atwood said.[149] "I agree that there are more risks in a lunar-orbit mission than in an Earth-orbit mission," Bergen said. "But I am confident that our systems will perform satisfactorily. Although there will be less risks with a circumlunar flight, there are risks with no gain. We should make the lunar-orbit flight."[150] It would be the flight with the most operational gain.

The next day, Mueller kicked off a meeting that included the center directors—Phillips, Paine, Newell, Shapley, and Finger—and a larger number of staff members.[151] Low, Kraft, Slayton, and Phillips gave presentations similar to the ones they gave the day before. Then Mueller updated everyone on the positive view of the mission from STAC (Scientific and Technology Advisory Committee), PSAC (President's Science Advisory Committee), the Department of Defense, and the Apollo executives. The only objection from these groups, Mueller pointed out, was from Gordon MacDonald from PSAC; MacDonald believed the risks far outweighed the benefits. Mueller also stated that Bellcomm, formerly of Bell Labs, was also negative on the risk-reward potential and felt the mission was not justified. Mueller then said that he had tried to perform a numerical risk analysis himself but that this had "not turned out to be as positive as he had hoped it would be."[152] Both Gilruth and Paine objected to Mueller's assessment. Paine then held a number of smaller sessions for executives, including Mueller, in which they discussed the mission at length. In all, there were three smaller meetings in which Paine had to overcome Muller's objections. "At the conclusion of these meetings," Low wrote, "Dr. Paine announced that the *Apollo 8* flight would be a lunar orbit mission. This was announced publicly in a press conference in Washington on Tuesday, 12 November 1968."[153]

"We got to the moon despite George Mueller," Frank Borman said, reflecting on the delicate negotiations in Washington. It was a sentiment felt by many who were aware of Mueller's objections to the flight. "We got to the moon because of George Low."[154]

As everyone in the meetings knew, despite the go decision, risks still remained to be managed. There would be no rest from the crushing demands of paying attention to details. Low would not allow it. The team was managing the downstream developments for *Apollos 9, 10,* and *11*; simultaneously, they were also juggling last-minute glitches with CSM-103. Five days before launch, a new problem arose around the hundreds of toggle switches within the CSM. "On the toggle switches, we now have 100% confirmation that all contact buttons are in place on the two CM/SM separation switches," Low wrote to Gilruth. "However, we discovered a new toggle switch problem. This was identified on the CM/SM separation switches. . . . In x-raying these switches, we found a small loose wire emanating from the central terminal and going toward one of the outside terminals."[155] Once again, short circuits within the toggles became a worry and concern.

"It is the small detail that counts," Low said. "Although there was only a one-in-a-million chance that a switch installed in Borman's spacecraft would fail, we had to do something about it. Today, I could tell you all about these toggle switches: how they are made, who makes them, and how they function. After learning all about them, we were able to devise a test to prove that the switches installed in *Apollo 8* were all right, and we were able to fly on schedule. A small switch—a one-in-a-million chance—yet most important to the guys who fly the spacecraft. Attention to detail."[156]

Low and his team had noticed short problems in a number of the toggles in the weeks prior, as well as some serious manufacturing concerns. These toggles were problematic not only in the CSM but also in the LM. While some toggles were prone to short-circuit, others were put together incorrectly and operated inversely to what the switch labeling indicated. Some switches might be in the on position, for example, when it looked like they were switched to off. X-rays of the switches over the weeks identified the problems. And all switches had to be examined. Since there were so many, Low decided not to take any chances. He worked with flight crew operations and the astronauts to make sure they had procedures in place should an unexpected switch problem occur in flight. This was especially true for those switches that were needed during critical maneuvers, like separation or orbital insertion. Low would spend an inordinate amount of time on the toggle switch issue. He would not allow the flight to proceed unless he was certain that the problem toggles were identified and replaced or that proper work-around procedures were established. "I spent all day on Command Module toggle switches," Low wrote to Gilruth. "The subject got to be so complex that I wrote a separate memorandum on it. . . . We are now at the point where I believe everybody is satisfied that we have done the right thing."[157] Still, throughout the entire flight of *Apollo 8*—including the all-critical burn of the main engine to escape lunar orbit, and the separation from the service module to enable the command module to reenter the earth's atmosphere—Low would be monitoring for any signs of problems or anomalies.

As for the new problem? "I worked with Borman on some last-minute procedural changes," Low wrote. "These included the changes to take care of the possible malfunctioning switches . . . and changes to take care of the newly defined engine start problem." For the newly discovered issue in the command module–service module separation toggle, they added yet another backup procedure: "We now have developed procedures that completely work around any possible inadvertent switch closure in the CM/SM sep area."[158]

Apollo 8 launched on Friday, 21 December 1968. Low was of course at the Cape for the launch, as he always was, and then returned to Houston to monitor progress from the back of mission control. As usual, he was seated with Bob Gilruth, Chris Kraft, and Deke Slayton throughout the entire flight. From the second the crew launched until the second they splashed down in the Pacific, Low continued to obsess over the details.

Sometimes Low would be asked questions, or he would get involved in decisions or analysis midflight. Other times, he would take part in press conferences or visit VIP guests. More importantly, Low would often interact with the family members of the astronauts—keeping them informed, keeping them encouraged and positive. Valerie Anders cherishes and remembers Low's presence during *Apollo 8*. When she and Low chatted on Monday, 23 December, one day before the crew reached the moon, she recalls sensing that her presence made Low nervous—as if he bore the weight of the burden that if something went wrong in the flight, he would be the one to make her a widow and her children fatherless. Robert Zimmerman writes in *Genesis: The Story of* Apollo 8, "His concern touched her, and Valerie tried to ease his mind.... She had faith in George Low, and she wanted to show him that."[159]

The same was true on Christmas Day 1968. *Apollo 8* had made a successful orbital insertion on the day prior. When telemetry reached Earth that the CSM was in orbit, elation broke out in mission control. "Kraft, Gilruth and Low, on the back row of the control center, could hardly contain themselves," Kranz said.[160] The crew would later captivate the world with a live television transmission of their descriptions of the lunar surface and with an emotional reading of Genesis while in orbit.

Now, it was time to depart. The capsule had traveled to the far side of the moon, out of communication with Earth. All that remained was for the crew to ignite the CSM engine with one of those potentially tricky toggle switches. Would the switch work? Would the engine refire after being cold for so long? These and other questions clearly filled Low's mind. Zimmerman writes, "This was by far his most fearful moment during the whole race to the moon. He hadn't been able to hide his worry in his conversation with Valerie Anders on Monday.... As confident as he had been that everything would work, he was about to find out whether he had been right, or whether his decision had caused the death of three brave men."[161]

Everyone waited tensely in mission control for the reacquisition of a signal from the capsule. Seconds turned to minutes. And then, over the crackling squawk box, came a response. Not a scream of "Fire!" like during *Apollo 1*, but the reassuring sound of Jim Lovell's voice: "Houston, *Apollo 8*. Please be informed, there is a Santa Claus." The switch and engine had worked. *Apollo 8* was heading home, triumphantly.

In the 3 January 1969 issue of *Time* magazine, Low was heralded as "The Groundling Who Won," in the aftermath of the wildly successful *Apollo 8* mission to the moon. Time quoted several NASA high-ranking officials as proclaiming that if it had not been for Low, there would not have been an *Apollo 8* flight to the moon.

Low was reportedly "as thrilled as any of his five children" by the flight. "I looked at the moon after *Apollo 8* went into orbit," he said. "It looked different."[162] And it was. The moon was no longer just a mysterious orb floating in the sky out of the reach of humanity, as it had been perceived for eons. Jules Verne's 1865 fantasy of three men going to the moon in a ballistic capsule shot into the sky from Florida was no longer science fiction; it was now science fact. Humans, for the first time, were cradled in its gravitational embrace. Kennedy's seemingly impossible goal of landing a man on the moon and bringing him safely home was now tantalizingly close to fruition. What some people within NASA had called a reckless and unnecessary risk had suddenly turned into as audacious of a success as any mission before, or after, *Apollo 8*.

Low considered *Apollo 8* the emotional highlight of the program.[163] For many years afterward, he would often gift the Bill Anders *Earthrise* photo taken from lunar orbit as a memento to friends, colleagues, and well-wishers. It was one of his favorite images. It perfectly captured the emotional payoff of one of the boldest risk-reward decisions of his entire career. Ten years later he gave a speech on productivity and innovation in America. U.S. productivity was faltering in 1979, and Low pointed out that productivity gains in the country were half of what they were between 1954 and 1966. A big reason for this decline, Low said, was a lack of innovation, which had been choked out by excess regulation driven by risk aversion. In the speech, Low went back to his experience on *Apollo 8* to make his point. In so doing, he put *Apollo 8* into a broader, national perspective.

In his mind, making these risk-reward decisions is the key to future human success on the planet. At a joint meeting of the Capital District Chambers of Commerce, held in Troy, New York, on 19 April 1979, Low said,

> New rules and regulations . . . are generally designed to protect us against the risks of life, which fail to recognize that without risk, there can be no gains.
>
> Let me digress and tell you an anecdote about Apollo. The time was ten years ago, and NASA was making preparations for one of its most daring ideas: *Apollo 8*, a flight to lunar orbit, a flight which had been conceived a scant three months earlier. . . . Just picture for a moment the situation which the engineers, program managers, and administrators of NASA then faced. Less than two years earlier, a tragic fire on the launchpad had taken three lives, and we were still recovering from the shock of that event. The Apollo spacecraft had been redesigned and rebuilt. It had just completed its first flight in Earth orbit. The Saturn V, which was needed for *Apollo 8*, had flown twice before, unmanned, but on its second flight, three of its hydrogen-fueled engines had failed. Nevertheless, we decided that in another month or so we should leave the confines and the relative safety of Earth orbit and send three men to the moon.
>
> Why?
>
> Because we believed the gains were worth the risks. I'll never forget how later on, in mission control, we were all acutely aware of those risks when the Apollo spacecraft with three men in it for the first time disappeared behind the moon—something that had never happened before. . . . This balance of risk and gain applies not only to Apollo. It applies to energy, to the environment, to the economy—in fact, it applies to our future as a nation, and perhaps even to our future as a human species . . .
>
> I hope my message is clear. Much needs to be done to rekindle the American spirit of innovation, and that spirit is needed to help keep America great. We must be leaders in technology, not followers—innovators, not imitators. It is up to us to be informed and to inform; to speak out and to respond to irresponsible arguments; and to let our elected repre-

sentatives in Washington know that the pendulum has swung too far. If we don't, nobody will.[164]

***

Just a few weeks after his forty-third birthday, George Low stood on the red, elevated gantry at launchpad 39A at Cape Kennedy Space Center in Florida late on the night before *Apollo 11*'s historic launch.[165] During the day, the temperature had been in the high eighties, but now it was a relatively cool and breezy seventy-nine degrees. Earlier that evening, the ground crew had finished fueling the massive rocket. It was now bathed in bright, dramatic floodlights.

He had chaired the NASA committee that recommended a manned lunar landing and provided the background work for President Kennedy's decision to go to the moon. And now, nine and a half years later, millions of visitors and anxious spectators were already gathered around the Cape and along the roads for the morning launch, which was scheduled for 9:32 a.m. on 16 July 1969. In the years preceding his work on the lunar landing goal, as head of manned spaceflight during the early years of NASA, Low had not been involved in the day-to-day engineering of Apollo until April 1967, a few short months after the *Apollo 1* fire.

Now, twenty-seven months later, he had not only set the grand plan in motion, but he had also come to know and understand, as no one else, two of the most complex flying machines ever built—the Apollo CSM and LM— via his role as Apollo program manager. Not only had Low helped to create this audacious and seemingly impossible dream; he was also one of the principle reasons why that dream was about to become a reality. Poised to unleash in just a few short hours a massive 7.5 million pounds of thrust, the mighty Saturn V rocket—accompanied by the complex flying machines Low helped rebuild—was about to carry Neil Armstrong, Buzz Aldrin, and Michael Collins into the history books.

Looking out in all directions from the highest peak on the coast, only about 1 percent of the new moon would be visible in the night sky. It might have struck Low as symbolic; that was probably about all we knew about the moon, in percentage terms, compared to the science the Apollo program promised to reveal. Right now, however, symbolism of another kind was on his mind. "There was no science—at least, no major science—on *Apollo 11*," Low said.

"That was to come later, starting with *Apollo 12*. They did bring back samples from the first flight, and they did leave some instruments on the moon—but the enormous amount of science that was to come back, came back from the later flights."[166] This flight, the first flight to attempt to land a man on the moon and bring him home safely, was as much about pure engineering as it was about science. More than anything, this was a symbolic mission of American ingenuity and technical achievement.

With him on the gantry was Jack Kinzler, a NASA engineer who got his start at NACA Lewis a few years before Low and who was now the chief of the Technical Services Center at MSC.[167] Kinzler was known as Mr. Fix It, for his many unique and important contributions to problem solving at NASA. One of the jobs he undertook was to procure a U.S. flag and build a flagpole system—which he loosely based on a hidden-sleeve and curtain rod system his mother used at home when he was growing up—that would later be used on all the lunar landings. Kinzler was charged with coming up with this particular flag system by a committee that explored what sort of symbolic items should be left by the *Apollo 11* crew on the moon. They had come up with the idea of planting a flag and putting a commemorative "We Came in Peace" plaque on the lunar surface. This committee was run by Willis Shapley, who came up with the plaque's general appearance as well as the original words. Gilruth sketched the two hemispheres, and he designed the plaque's shape. Of course, the committee got politicized, and all kinds of suggestions were being forced on it from all directions.

Low was rather indifferent to it all, until he heard directly from Washington that an effort was underway to push for planting a UN flag instead of the Stars and Stripes. "The only strong position I took in all of this from the very beginning was that we should display the United States flag and not the United Nations flag," he said. "In fact, when Sam Philips first sent me a note . . . which pointed toward the U.N. flag, I called him back and told him, 'Over my dead body!'"[168]

Now, the night before the launch, yet another of the millions of infinitesimal details going into the *Apollo 11* mission was not too small for him to obsess over or monitor. He carefully watched as Kinzler used metal latches to attach the commemorative plaque to the center leg of the LM, and then stowed the flag for easy access by Neil Armstrong and Buzz Aldrin. While the world would be watching simultaneously in one of history's most famous, global communal events, Low took the extra effort to make sure it would be recognized as an U.S. achievement as Kennedy had intended. The goal was achieved. The moon race, which Low had helped define back in 1961, had been won.

Throughout the preparation and lead-up to the flight, Low stayed close to the *Apollo 11* crew, as he had with all the preceding crews. They were as much the designers and engineers of their mission as Low, Kraft, Slayton, or anyone else. "I spent considerable time today with the newly named *Apollo 11* crew in talking to them about all of the things that might be of concern at the press conference," he noted for example in January 1969. "We touched all sorts of subjects from the probability of *Apollo 11* actually being a G mission [lunar landing] to weights, delivery schedules, lunar surface timelines, the requirement of the F mission, etc. I think we got a real good crew."[169]

In addition to getting a really good crew, they got a great commander in Neil Armstrong. Low's experience with Armstrong extended from first meeting him at NACA Lewis to going with him and Dick Gordon on their South American tour during the Gemini program. He was impressed with Armstrong's cool handling of his near-fatal Gemini flight. He also watched and studied Armstrong throughout his training and preparation for flight. To say he was impressed with Armstrong's capabilities would be an understatement. He could think of no one better not only to command the mission but also to be the first man to walk on the moon. "Neil Armstrong had the same qualities as George Low," Gerry Griffin recalled, comparing both men and noting the admiration each had for the other. Both were cool in emotional situations; both were not prone to overreact; both were not big talkers but great listeners; and both could take in a multitude of rapid-fire inputs at once and quickly and calmly assess a situation to make the right decision. They were both what Griffin called "special talents."[170] When Gilruth approached Low for his input on the first-man decision—between Aldrin and Armstrong—he was in agreement. Three months before launch, Low announced that Armstrong would be first man.[171]

Low's involvement with the crew, as well as his constant asking of questions, sometimes would get misinterpreted by those in Washington. He at one point asked Armstrong if he had thought about what he would say as his first words on the moon. Armstrong simply replied, "Yes."[172] He also talked to Si Bourgin about symbolic items for Shapley's committee. Low respected Bourgin, a former *Newsweek* bureau chief and a science policy adviser at the U.S. Information Agency.[173] It was Bourgin, after all, who worked with Borman on the famous Christmas Eve reading of a biblical passage from Genesis; might he have similar ideas for *Apollo 11*?

When Julian Scheer, the head of NASA Public Affairs in Washington, caught wind of what he mistakenly thought was Low trying to influence Armstrong's first words from the moon, he went ballistic. Without calling Low or confirming the rumor, Scheer fired off a three-page lecturing memo on the need to let the astronauts speak for themselves and in their own words. He quoted everyone from Columbus to Capt. James Cook, Meriwether Lewis, and Robert Peary—just in case Low did not pick up on his not-so-subtle point.[174] Clearly, Scheer did not know Low well enough, or his views on this matter.

When Frank Borman and the crew read from Genesis on *Apollo 8*, NASA received a considerable backlash from certain circles that believed religion should not be supported by a tax-payer-funded entity like NASA. Madilyn Murray O'Hare famously brought suit against NASA in this regard. (She lost.) NASA also received a huge outpouring from people who supported the right of the astronauts to freely express themselves. As program manager, Low received a lot of letters, for and against. In response to many of these letters, he let his views be known that he fully endorsed the reading: "It has always been our policy to let our astronauts think and speak for themselves, and it is entirely up to them as to how they are best able to express their thoughts." He even went on to quote the prayer that Gordon Cooper said on the last flight of Mercury, underscoring Cooper's asking of God not only to bless the mission and the people working on it but also to "help us in our future space endeavors that we may show the world that a democracy really can compete, and still are able to do the things in a big way," while imploring people to "help one another, and to work with one another rather than fighting and bickering."[175]

In response to Scheer's memo, Low wrote back in a more subtle manner to explain that there had been, obviously, a misunderstanding on Scheer's part. He further clarified that he had, indeed, spoken with Simon "Si" Bourgin but that their conversation was about what the astronauts should *do* (not *say*) when they first landed. "It is properly NASA's function to plan what artifacts should be left on the lunar surface, or what should be brought back, but the words that the astronauts should say, must be entirely their own," he wrote Scheer. "I have had a meeting with Neil Armstrong to discuss with him some of our ideas and suggestions, including those of Si Bourgin's, in order to solicit his views. Even though I had not yet received your letter at that time, we also discussed the point that whatever things are left on the lunar surface are things that he must be comfortable with, and whatever words are said, must be his own words."[176] As history would record, no one ever told the astronauts what to say. Such freedom is at the heart of a democracy—and an integral part of the program, from Low's perspective. The astronauts' words were entirely their own.

Traffic in and around MSC was just as thick and crowded as at the Cape for the launch. Still, while driving slowly in for the first day at mission control, Low noticed Joe Shea walking toward the center. Shea did not have a car, or he was just being smart with all the traffic. Either way, Low pulled over, gave him a ride, and proudly walked into the building with him. As much as the *Apollo 1* fire was everyone's, so, too, *Apollo 11*'s triumph would be everyone's success.

During the mission, Low took his post in the back row of mission control, as always, helping out wherever he could. "He had an official seat at my console in the back row of mission control," Kraft explained.[177] "Every top manager from NASA was in the glass-walled viewing room." Kraft felt that Gilruth and Low deserved to be in the middle of the action and not walled off in the VIP viewing room. They were a team. When *Apollo 11* actually landed on the moon, Kraft recalls, Low actually tried to hug him. "He never showed that kind of emotion," Kraft chuckled. But Kraft's headset cords got in the way, so the men settled for a hearty handshake. Later, they would celebrate with a drink in Kraft's office. "Low's discipline extended to alcohol," Jerry Bostick said, in recalling a story that Kraft told him. Sure, they smoked victory cigars at all the missions, but alcohol was a different matter. The story may be apocryphal, but it is nonetheless illustrative of the emotion that broke free at the moment of the landing. "After he was named Apollo program manager, Low vowed not to drink any alcohol until we had successfully landed on the moon. As luck would have it, I was assigned the office space which was Kraft's back in 1969 when we landed on the moon. I noticed that one of the ceiling tiles had a dent in it. Kraft told me that's where he and George Low had uncorked a bottle of champagne to celebrate, and the cork hit the ceiling."[178]

"To me, *Apollo 11* was the culmination of two and a half years of the most intensive period of work that I have ever been involved in," Low said.[179] At the time, and in the moment, there was not a lot of philosophical reflection on the significance or meaning of the event by people in the program. Sure, the media had a field day, exploring the actual and philosophical importance of it. But for Low and his team, the time had been a whirlwind, and there were still several important missions coming up in rapid succession that needed their full attention.

On the occasion of *Apollo 11*'s fifth anniversary, Low discussed how the mission's significance was starting to come into focus for him. If he had been asked immediately after the mission, as a man deep in the trenches—he probably would have talked about the meaning in terms of the development of technological and operational capabilities. As it was, in hindsight, he drew a much broader and deeper meaning:

> There are many facets, many things one can think back about and relate Apollo to now, to what it really meant. I think first of all, it gave something to ourselves, to the United States and to the World, something that we can indeed be proud of. If you think back to the period of the 1960s, we weren't very happy ladies and gentlemen. We had the ever-deepening involvement in the Vietnam War, we had riots on the campuses, riots in Watts and elsewhere, three terrible assassinations, a great deal of strife and turmoil in the country. And yet, all of these then were perhaps overcome, if you will, by one single event, and that was *Apollo 11*. The decade ended with it. It was a fantastic adventure that I believe helped overcome all of the bad things of the decade—to represent Americans, to ourselves and particularly to the rest of the world, as we like to see ourselves, and as we hope the rest of the world likes to see us.[180]

While the rest of the nation celebrated *Apollo 11*'s grand success, Low remained in Houston, as preparations for *Apollo 12* were gearing up. That mission was set to fly in just a couple of short months. There was a lot of important work to get done. Once again, lives depended on a complete focus and attention to detail. He wasn't sure it would be fair to future crews to continue on as program manager. He had the capability to do it, but he wasn't so sure he still had the drive. It demanded 100 percent focus and concentration. It was time for Low to move on. From his perspective, he had fulfilled the job he was asked to take at Washington National Airport in April 1967. He got Apollo back on track. He got the nation to the moon. Now, he wanted to return to deputy center director and get more of his life back.

I felt when I took the Apollo job that I had committed for the duration. I felt that once we had landed on the moon, I had finished that commitment. I am sure I could have continued with the job at the same pace. I was handling it. I never felt at any time that I was burnt out or any of the things that you hear about. But once we had done it, I don't think I could have gone through another flight in that capacity. I didn't think it would be fair to the guys flying on the next flight. I just didn't quite have the same drive to really worry about all the things that one had to worry about—the infinite and infinitesimal detail.[181]

At about the same time, NASA had a new—perhaps even bigger—problem. Rather than facing fallout from a failing program, NASA was facing an existential crisis. In successfully landing a man on the moon before the end of the decade, NASA had become the rarest of all government agencies: it actually achieved what it seemingly was established to do. Now, the knives were out. NASA's peak budget was in 1966—three years before the moon landing on *Apollo 11* and a year before the *Apollo 1* fire required a lot of expensive, unbudgeted reengineering. Even before *Apollo 11* lifted off, Houston and other Apollo centers were already experiencing mandatory reduction-in-force layoffs. "Shortly after the *Apollo 7* landing, Congress announced there would be no space program beyond the Apollo Application Program, a planned mini space station—dubbed *Skylab*—that would use leftover Apollo hardware," Kranz said. Several later Apollo missions were already being scrubbed for budgetary reasons. "The space program was suffering. The lunar program was coming to an end. I felt betrayed," said Kranz, expressing the opinion of many who worked in the trenches during Apollo.[182]

But not everyone felt as Kranz did. Internally, the various NASA centers felt that Apollo was sucking up too much of the budget money and agency priority. Specifically, the unmanned space program centers wanted their shot at a series of ambitious projects, and so did the earth science areas. Mueller and his Advanced Applications Program had floundered, as Low had predicted it would, and Mueller would leave NASA before the year was out. President Nixon—cognizant that Apollo would always be viewed as John F. Kennedy's program and achievement—was more interested in jobs than he was in establishing another big, bold (not to mention expensive) space program.

The Sunday after *Apollo 11* splashed down, Low planned to get some work done in the office and give a speech early in the week, and then he and Mary R. planned to get away for a few days. After returning to the office that Monday, as he sat in his black tie and tux at his desk, getting ready to give a speech at a formal celebration in the city, he got an unexpected call from Paine. "How would you like to come to Washington to be my deputy?" he was asked. A bit taken off guard, he wanted a little time to think about it. "Well, as long as you are going to think about it," Paine said, "I'm going to send your name to the president for consideration anyway." Rather than fight it, Low figured his nomination would be a long shot anyway; he imagined the White House would select someone else on Paine's list over him, probably someone more politically connected with the current Nixon administration.

Low waited until he and Mary R. were on a plane to the Bahamas before broaching the subject. "I really didn't want to leave Houston," he recalled later. "We had decided to stay there forever. I had planned on going back as Bob's deputy and eventually getting the center. I thought that would be a very nice kind of family life to lead for the next several years."[183] As Low and Mary R. discussed it, he thought it over. He never expected to get beyond a section head at the NACA, let alone do the things he had done. It was an honor to be offered the job as deputy administrator. The administrator is a political appointment and deals with the president, the White House staff, and Congress. The deputy administrator, on the other hand, is really Mr. NASA, the internal career person who focuses on the agency's day-to-day activities. In the business world, the administrator would be analogous to the chairman of the board; the deputy administrator, to the CEO. Low's "career in reverse" was about to catapult him to the top of his field. Not a bad result, even if it was a long shot.

"Again, the experience as deputy director in Houston was very, very good for me, and I hope somewhat good for Houston, but I can't judge that," Low said. "I learned a great deal, and I also made mistakes. But I hope I brought a little bit to the organization. Everything at the NACA or NASA somebody said, 'Hey, we want you to do this.' I never went looking for a job. Before I was ready to look for things, somebody said, 'Let's put him somewhere else.'"[184]

If NASA needed George Low in Washington again, who was he to say no?

ANGERN, N.Ö.

J.P.W.

1. Angern Castle, the Low family home in the 1920–30s. Courtesy of the Low family.

2. Low as a young boy in Austria. Courtesy of the Low family.

3. Low playing with his beloved motorized erector set as a child in Austria circa 1932–33. Courtesy of the Low family.

4. Low's father, Arthur, who died in Austria of cancer at the age of thirty-three in 1934. Courtesy of the Low family.

5. Low's mother, Gertrude. Courtesy of the Low family.

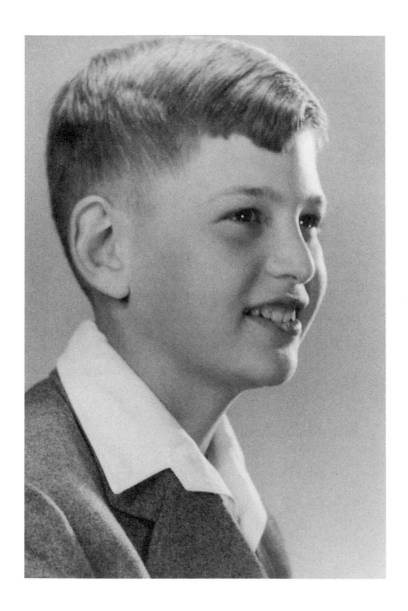

6. Low at age thirteen, the year he immigrated to the United States. Courtesy of the Low family.

7. Low in his military uniform in the 1940s. Courtesy of the Low family.

8. The Low family farm in Milford, New York, circa 1950s. Courtesy of the Low family.

9. Low and Mary R. in Cleveland in 1952. Photo by Bob Blue; courtesy of the Low family.

10. (*opposite top*) Low teaching an advanced course in aeronautics at NASA
Lewis Research Laboratory in Cleveland in 1954. Courtesy of NASA.

11. (*opposite bottom*) Low teaching a course on orbital mechanics at NASA
Lewis Research Laboratory in Cleveland in 1958. Courtesy of NASA.

12. (*above*) Low with President John F. Kennedy, along with astronauts Gordon Cooper
and Virgil "Gus" Grissom in front of a Gemini capsule at a press conference at
Cape Canaveral, 16 November 1963. Courtesy of NASA.

13. Low conducting a press conference during Project Gemini in
Houston, Texas, 16 November 1965. Courtesy of NASA.

14. Chris Kraft, Bob Gilruth, and Low conferring in the back row in mission
control in Houston, Texas, during Project Gemini in 1965. Courtesy of NASA.

15. Chris Kraft and Low in mission control in Houston as they discuss and celebrate the successful splashdown of the *Gemini 11* spacecraft. Courtesy of NASA.

16. (*above*) Low (*second from left*) talking with Gen. John P. McConnell (*right*) during a 1966 visit to the MSC, Houston. *Left to right*: Lt. Gen. Leighton I. Davis, George M. Low, Robert R. Gilruth, and Gen. John P. McConnell. Courtesy of NASA.

17. (*opposite top*) Low and Bob Gilruth inspecting the returned *Apollo 7* spacecraft in 1968. Courtesy of NASA.

18. (*opposite bottom*) Low, Glynn Lunney, and Chris Kraft discussing *Apollo 8*'s progress during its return flight on 26 December 1968. Courtesy of NASA.

19. Low and Chris Kraft congratulating each other on a successful splashdown of *Apollo 8* on 27 December 1968. Courtesy of NASA.

20. Low conducting a meeting in his office as Apollo program manager in 1969. Courtesy of the Institute Archives and Special Collections, Rensselaer Polytechnic Institute, Troy, New York.

21. Low intently monitoring progress in mission control during the *Apollo 11* lunar landing on 20 July 1969. Courtesy of NASA.

22. Low expressing his delight in mission control during the first steps on the moon by Neil Armstrong on 20 July 1969. Courtesy of NASA.

23. (*opposite top*) Chris Kraft, Low, and Bob Gilruth in mission control
during the *Apollo 11* splashdown on 27 July 1969. Courtesy of NASA.

24. (*opposite bottom*) Low smiling broadly while accompanying the first *Apollo 11* sample
return container with lunar surface material at Ellington Air Force Base, 25 July 1969.
Happily posing for photographs with the rock box are (*left to right*) George M. Low, Lt. Gen.
Samuel C. Phillips, George S. Trimble (*almost obscured*), Eugene G. Edmonds, Richard S.
Johnston (*in back*), Thomas O. Paine, and Robert R. Gilruth. Courtesy of NASA.

25. (*above*) Low at his desk at the MSC in Houston in 1969. Courtesy of the Institute
Archives and Special Collections, Rensselaer Polytechnic Institute, Troy, New York.

26. Low (*right, with cigar*), Thomas Paine (*center of frame*), and other NASA officials applauding the successful splashdown of the *Apollo 13* mission in the MSC mission controlcenter, located in building 30. *Apollo 13* splashed down at 12:07:44 p.m. (CST), 17 April 1970, in the South Pacific Ocean. Courtesy of NASA.

27. Low wearing his T-38 flight jacket during one of his many NASA facility tour visits in the 1970s. Courtesy of the Institute Archives and Special Collections, Rensselaer Polytechnic Institute, Troy, New York.

28. Low and other NASA officials gathering around a console in the Mission Operations Control Room (MOCR) in the Mission Control Center prior to deciding to land *Apollo 16* on the moon (rather than abort the landing due to oscillations of the spacecraft). Seated are (*left to right*) Christopher C. Kraft Jr. and Brig. Gen. James A. McDivitt; standing are (*left to right*) Rocco A. Petrone, Capt. John K. Holcomb, Sigurd A. Sjoberg, Capt. Chester M. Lee, Dale D. Myers, and George M. Low. Courtesy of NASA.

29. Low discussing the Apollo-Soyuz Test Project mission at a press conference in Houston with Boris N. Petrov, a senior academician and Apollo-Soyuz Test Project negotiator at the Soviet Academy of Sciences, on 13 July 1972. Courtesy of NASA.

30. (*opposite top*) Low bringing the three prime U.S. astronauts and Soviet commander and officials of the joint U.S.-USSR Apollo-Soyuz Test Project mission to visit President Gerald R. Ford in the White House on 7 September 1974. *Left to right*: Vladimir A. Shatalov, Valeriy N. Kubasov, Aleksey A. Leonov, Anatoliy Dobrynin, the chief executive, George M. Low, Thomas P. Stafford, Donald K. Slayton, and Vance D. Brand. Courtesy of NASA.

31. (*opposite bottom*) Low talking with Russian Apollo-Soyuz Test Project team cosmonauts Romanenko (*center*) and Ivanchenkov (*right*) in Washington. Courtesy of NASA.

32. (*above*) A reflective Low in mission control in Houston during the last Apollo flight, Apollo-Soyuz Test Project, in 1975. Courtesy of the Institute Archives and Special Collections, Rensselaer Polytechnic Institute, Troy, New York.

33. George M. Low RPI presidential portrait, 1980s. Courtesy of the Institute Archives and Special Collections, Rensselaer Polytechnic Institute, Troy, New York.

34. Low passing out doughnuts to students sleeping in line for RPI hockey season tickets. Courtesy of the Low family.

35. (*above*) Low working on a speech in his den in Troy, New York, in the 1980s, a green felt-tip pen in hand and an *Apollo 9* signed photo on his wall. Courtesy of the Low family.

36. (*opposite top*) Low and his wife, Mary R., in 1982 at home in Troy, New York. Courtesy of the Low family.

37. (*opposite bottom*) Mary R. accepting Low's posthumous Presidential Medal of Freedom from President Ronald Reagan at the White House on 23 May 1985. Courtesy of the Low family.

To Mary Low
With best wishes, Ronald Reagan

38. The Low family at the dedication of the George M. Low Gallery Center for Industrial Innovation on 27 September 2002. *Left to right*: Mark Low, Nancy Sullivan, John Low, Mary R. Low, Diane Murphy, and G. David Low. Courtesy of the Low family.

5

Post-Apollo

At the dawn of the 1970s, grand space expenditures were viewed as a legacy of the Kennedy and Johnson administrations; for Nixon's White House, and for the rest of America, audacious initiatives such as Apollo represented nothing more than expensive bills that never seemed to get paid. According to a poll in *Newsweek*'s "Troubled American" issue, more than 56 percent of the country wanted the government to spend considerably *less* money on the space programs.[1] In the words of *New York Times* writer John Noble Wilfred, neither the government "nor the American public seems in any mood to pledge the money for another accelerated, Apollo-like space project."[2] In a very short period of time, NASA had gone from national hero to national zero. It was hard for people in Houston to wrap their heads around it at the time, yet it was easy to feel and experience the mood if inside the DC Beltway.

Low arrived in Washington for his prenomination interviews on 15 October 1969. On his way to the White House, he had to make his way through the crowds of Moratorium Day marchers—a protest march against the Vietnam War. There was no doubt in his mind that a page had been turned in American society. In the old Eisenhower Executive Office Building on Pennsylvania Avenue, he met with Lee DuBridge, the president's science advisor and a frequent critic of NASA human spaceflight. DuBridge was surprised to see him and wasn't prepared for an interview, having thought that Low was already on the job.[3] They made awkward small talk, then Lee got right down to business. With the din of counterculture protesters outside, he explained his two pet peeves concerning NASA: the lack of emphasis on science over engineering within the agency and the current overweighting of expensive and risky human spaceflight programs over less costly, more efficient robotic missions.

"It was really a culture shock," senior NASA strategist Willis Shapley recalled of the change in the national and political mood toward human spaceflight during the 1970s.[4] Coming from MSC in Houston—the hub of all U.S. human spaceflight activities—it was equally so for Low, but he got over it quickly. It wasn't the first time he had experienced the jarring cognitive dissonance between NASA Headquarters and the field centers on big policy issues. The view from Washington was rarely the same as the view from Cleveland, or Houston, or any of the other NASA installations. He'd been experiencing that disconnect since he left Lewis to come to Washington in the late 1950s. In fact, the experience informed his career-development philosophy, which argued that rotating through NASA Headquarters was critical in a future NASA leader's career.

While not a rousing welcome from DuBridge, it was perhaps better than the impression Low was about to get from his next meeting. "First, the guard at the White House gate had no record of my appointment," he explained of the awkward encounter.[5] "After some discussion and phone calling, he sent me to a guard in the basement of the West Wing. Once I arrived, the guard said, 'You can leave it right here.'" Apparently, the guard thought Low was a delivery boy. After the exchange of a few brief words, he was asked to wait. Sitting on a bench in the basement, Low watched as large groups of people came and went. In between, he casually read a two-day-old newspaper. Welcome back to Washington, Mr. Low.

After a prolonged period of time, a staff assistant suddenly appeared and ushered him into a White House conference room. The appointment was with Pete Flanigan, an influential assistant to President Nixon who was NASA's main point of contact. Flanigan's current opinion of NASA was less than flattering. Without wasting any time, Low and Flanigan chatted a bit about the post-Apollo STG report, recent lunar scientific results, and the current scientist-engineer debate. It was a terse exchange. There really wasn't time to go into any detail or substance. In Low's estimation, Flanigan was dismissive; so much so that Low described him as "quite provocative." In short order, Flanigan stood up. He announced an urgent meeting he had with the president. And he left. The interview was over. "I must not have been communicating very well," Low surmised, thinking, perhaps, he'd be heading back to Houston for good.[6]

On the contrary, Flanigan simply had heard all he needed to hear. He wrote Nixon a memo a few days later sizing up Low as "obviously a very capable individual." Flanigan had also done a reference check on him via one of Nixon's favorite astronauts, Frank Borman. Borman, Flanigan noted, had given his old boss "his complete support" and called him "perfectly capable of assuming the utmost responsibility."[7] The president quickly approved Low's nomination.

He cleared the Senate confirmation hearings on 25 November 1969, with several questions closely mirroring the issues raised by DuBridge. "I have always had very intense interest in all aspects of our exploration of space: in the applications effort, where we have learned so much, where we do so much in communications, in Earth resources, and in weather satellites; in the space science effort, where we have learned so much about the universe that is around us; and in manned space flight, of course," Low answered in response to Sen. Margaret Chase Smith, a NASA critic who had asked if he would have any problem balancing manned and unmanned flight priorities. "I do not foresee any difficulty in being able to properly supervise all of these areas and help in the management of all of them."[8]

Satisfied, the senator followed up on the topic of neglected science within NASA. How did Low see it, and what was he going to do about it? "This has been a very serious problem, and continues to be a problem, that deserves a great deal of attention by all of us," Low explained. "The problems stem from the fact that the initial goal of Apollo had to be basically an engineering effort. All of our attention had to be devoted to solving the engineering problems." Still, he pointed out, *Apollo 11* did accomplish some science, including installing a small measurement station on the lunar surface and bringing back forty-five pounds of lunar material for scientific analysis. Those samples were being shared with scientists around the world. He talked about increased scientific cooperation—and results—from *Apollo 12* in recent weeks and more science to come in future missions, especially the later Apollo missions. To increase the scientific effort, Low underscored that he has begun discussions with Dr. Paine about hiring a "chief scientist, whose function would be to strengthen NASA's relationship with the scientific community and strengthen the science program."[9]

With no more questions of substance from the assembled senators, James J. Gehrig, congressional staff director for the Committee on Aeronautical and Space Sciences, asked for Low's view on the need for greater international cooperation in space, a growing area of interest for the Nixon White House. "I have been quite interested and alert to the cooperation and participation with countries other than the United States, even with the Soviet Union,"

Low said. "I think this is a most important aspect of our space effort. I think it is one that in the second decade of space should be broadened and that we should seek the widest possible participation and cooperation that we can get."[10]

Low indicated that international cooperation on the Apollo missions would have been too complex because of the language barrier and technical-training consideration. He did emphasize, however, that he thought the era of orbital space stations would open up the possibility for more direct international participation. On the specific subject of Soviet cooperation, Low remained hopeful, though he cautioned that most direct efforts to date had proven futile. Still, he pointed out, the importance of a recent visit by several cosmonauts to the United States might signal a thawing in the relationship. "I think it was a very large step in getting our astronauts and their cosmonauts to talk together, to open up a free exchange. I think it is our job to continue to ask them to join with us in our space ventures. It is up to them then to respond; so far, they have not."[11]

No further questions were offered. Several senators—including Tom Dodd of Connecticut and Bill Saxbe of Ohio—called him "an excellent choice" and expressed their firm support for Low's nomination. Formally confirmed by the Senate, he was sworn in by Paine as deputy administrator on 3 December 1969. Utmost responsibilities, in Borman's terms, really did await him. In less than a year, Paine would resign as administrator, and Low would find himself helming NASA through perhaps one of its most uncertain and daunting periods, a time when factions in the White House would have liked nothing better than to defund NASA and split up its activities among other government agencies.

Thankfully, Low was on the side of those who felt that the 1969 post-Apollo STG long-range plan was too ambitious for the current climate. He was not coming to Washington seeking to push any grand, expensive schemes. Authored under Paine's leadership, the STG long-range plan caused agitation among White House staffers for its political and economic tone deafness.[12] The ambitious plans also bothered some members of the STG committee, like von Braun. It was hard to swallow a grand vision for interplanetary missions to Mars and beyond at a time when NASA was laying off staffers in large numbers across the country. As demand out of NASA declined, so, too, did employment within the aeronautical industry as a whole. "I have always felt that it would be a good idea to read the signs of the times and respond to what the country really wants, rather than try to cram a bill of goodies down somebody's throat for which the time is not ripe or ready," von Braun said in a defensive swipe at the report.[13]

Like Low, von Braun was politically astute enough to measure the prevailing winds in Washington. Paine's efforts were dead on arrival. Nixon didn't even bother to respond for more than six months after his plan was formally submitted. His problem? The price tag. Paine was looking out into the future for $5 billion to $7 billion development budgets, which were set to grow to over $10 billion to $15 billion in annual operating expenditures to support lunar bases, orbital space stations, manned missions to Mars, and an assortment of other space "goodies." Most people viewed those estimates to be wildly conservative since they were modeled on George Mueller's Integrated Space Transportation Plan of shuttles and space tugs—the same plan Low knew to be too expensive. The budget numbers for that plan were suspect because Mueller was "forcing people to give him numbers that were a lot lower in many areas than people wanted to give him."[14]

While Paine was asking for $5 billion, Flanigan and Nixon's budget people were talking about taking NASA in the opposite direction. Some members of the administration called for scaling back NASA largesse by more than two-thirds of its current budget, down to a mere $1.5 billion. Mars was a bust—that was a given. But with only $1.5 billion? So, too, human spaceflight. And most of robotic flight too.

Upon arriving in Washington, Low tried to be diplomatic and supportive with Paine. After all, Low was the new guy, and the administrator owned the vision. "I mentioned that I was concerned that we might be 'buying-in' to a program that was far more difficult than we now anticipated," he noted telling Paine.[15] He was also concerned with selling the shuttle based solely on Mueller's unrealistic cost-benefit tables. Instead, he preferred selling it on the system's capabilities. Spaceflight is expensive. He rightly feared that the low-cost approach would come back to haunt them. "I was particularly concerned that we might be selling it at too low a cost and would forever be defending cost increases. I mentioned that one of the things that made Apollo feasible was Jim Webb's original estimate of between $20 to $40 billion," he explained. Apollo came in at around $24 billion; in his original Low committee report, he had estimated the program at just $7 billion. Webb, an old budget hand, knew the game of government budgets. He multiplied Low's estimate by a healthy fudge factor. It worked. "With Webb's high estimate, we never had a problem in defending Apollo costs again during the early days of the program," he explained to Paine.[16]

"The agency's initial budget request in 1970 started out at $4.5 billion for 1971. It was based on moving out with speed on the Space Task Group's planning," Low remembered. "About the time I came in, we got one budget after another, and we had meeting after meeting, and each time, we came back with a little bit less money than before."[17] In the end, NASA's budget got slashed by a third, coming in at just a little over $3 billion. New program starts, like the human mission to Mars and an ambitious lunar base, as well as old programs, would have to be deferred or slashed, including later flights for Apollo.

Apollo was a ripe target in both Congress and the White House.

While Low conceded the need to eliminate some of the later flights—including *Apollos 18*, *19*, and *20*—he fought to draw the line at *Apollo 17*. He argued that the scientific payoff would be greater in the later flights than earlier. Cutting too many prematurely might also put the current flights in jeopardy because of low morale among the flight support staff. The situation on *Apollo 17* came to a head when George Schultz, Nixon's director of the Office of Management and Budget, called Low and explained that Nixon wanted it canceled because it was scheduled to occur in November of an election year. Nixon did not want to risk losing a flight during an election. "I indicated we must recognize every time we fly an Apollo mission, the risk is substantial," he wrote in his notes about his conversation with the White House.[18] "However, I pointed out that I did not believe that the risk of *Apollo 17* may be any higher than that for all other missions. I said that the risk to *Apollo 14*, *15*, and *16* would be great if we canceled *Apollo 17* because we would, in fact, be cutting the dog's tail off by inches and would not be able to maintain any stability in the Apollo program." Instead, he pragmatically suggested moving *Apollo 17* to December. This would serve a dual purpose. First, it would solve the president's election risk issue; second, the increase in time would ensure that all the scientific experiments were flight ready in time for launch. Nixon accepted the proposal. *Apollo 17* was moved to December. Low had saved the last Apollo lunar landing mission, making it available for them to select the only trained scientist to go to the moon: geologist Harrison (Jack) Schmitt.

Frustrated by his diminishing budgets and at dramatic odds with the White House over program priorities, Tom Paine would resign in September 1970, just eight months after Low started as deputy administrator. As late as 4 April 1970 Low held a meeting with Dr. Russell Drew of Nixon's Science and Technology Office. Drew told him that the Bureau of the Budget and the White House—particularly Flanigan—were very reluctant to give NASA the go-ahead on one of the last big-ticket items remaining from the STG plan—the shuttle. "We will have to do a great deal of work to convince them that we should move out" with the shuttle, he wrote in his notes for the record.[19] With Paine abruptly out of office, that work would now rest on Low's shoulders. At the age of forty-four, he was now NASA's acting administrator. For the time being, the agency's future viability rested with him.

"George is generally credited with being the guy who really brought Apollo back after the fire," remarked Francis Hoban, who worked as Low's assistant in 1970. "It was a tough road in those years. We had no similar support from the White House, Congress, or the public like we did during Apollo. The fact the agency still exists may be a tribute to George's ability to keep things going."[20] Paine had optimistically told Low that the president was committed to the shuttle and that he would be coming to Washington to build and not dismantle NASA. This turned out not to be the case. The situation with the budget and the economy were changing rapidly, and the budgetary process soon turned into a zero-sum, slippery-slope game of cuts and reductions. "President Nixon faced the job of balancing the budget, and he had a hell of a tough job in 1970," Low observed. "His general reaction to us was that he wanted to move forward at this stage with a meaningful space program. He did not want to be the president that stopped all the wonderful things that had happened. Still, he had to cut us back to a point below which we would have liked. We decided to support it as best we could, and to build the best program we could, out of the money we were given."[21] Low's philosophy was to strike a balance between the practical and the visionary and to move the agency forward by investing in future technologies and capabilities, rather than remaining invested in the old and outdated. "We are becoming introspective as a nation," he noted. Therefore, NASA needed to be, too. "We are asking ourselves, What does the country want? What can we do that's interesting with a reduced budget?"[22]

To support the effort, Low and his team would have to make tough compromises. It was a realpolitik that the Nixon administration wanted from Paine, but they didn't start to receive it until NASA was under Low's guidance for the nine months he served as acting administrator. Jim Fletcher, who would become administrator in May 1971, would further support this approach. Before Fletcher's appointment, however, Low's work would finally stop the budget bleeding. It would give Flanigan and the rest of the staff in the White House more confidence in the management of NASA and allow the agency to move the shuttle out of the planning and into the design phase. "The brakes have now been applied to NASA's skidding budget," he told reporters in March 1971, after spending the last six months working the problem. "A proposed $3.27 billion budget before Congress for the first time has halted the downward trend that has existed for four or five years, and we are requesting funding at exactly the same level as last year. This budget also halts the downward trend in employment. As you know, NASA has gone from a peak of 390,000 people in American industry down to about 108,000 at the end of this fiscal year. We will halt that downward trend with this budget."[23]

"We will move out with the 1972 budget with the programs we have designed for the 1970s." This transition from the 1960s to the 1970s "will allow us to proceed with exciting manned and unmanned programs," he said. Apollo was to maintain three more lunar flights, and the *Skylab* manned space station experiment was to be launched in 1973 and serve for several missions through 1974. "Then there will be a hiatus, a gap in manned spaceflight, until the shuttle comes along at the end of the decade. As we move out into the '70s, the space shuttle will bring together the best of the manned and unmanned programs to use a space transportation system that will put payloads into orbit in the most economical way."[24]

When Low first started presenting NASA's manned flight goals to Congress in the late 1950s, he emphasized the early program was just "the first step" in the country's exploration of space; in the future, he argued, there would be space stations and regular, reusable shuttles for low-cost and frequent trips into space. In a time when America was in a space race with the Soviets, a ballistic capsule spacecraft became the choice in Mercury, Gemini, and Apollo because it offered the least—albeit still formidable—technical development challenges and cost. With Low's experience at the NACA working on aerodynamic research for the X-15 and other winged aircraft at the edge of space, the shuttle itself had always appealed to him in theory as the next evolutionary step in technical capability. To advance the United States' engineering and technological base, a reusable, winged spacecraft like the x-15 would become his focus for human spaceflight.

"We spent many hours discussing and thinking about what might make sense for NASA—a shuttle without a space station, or a space station without a shuttle," he recalled. Even in his original Low committee report of 1961, he had highlighted that the long-term goal of developing an Earth-orbital space station was a logical progression from a lunar landing. However, "a space station without shuttle, without a good transportation system to build it and to make use of it, makes no sense at all. A shuttle without a space station does, and so we decided to go for the shuttle first. Get that development behind us, before we introduced a new, expensive start such as a space station."[25]

With NASA's robotic flight programs, the ambitious Grand Tour of the outer planets currently aimed to send out two robotic space probes to explore the outer planets: Saturn, Jupiter, Neptune, Uranus, and Pluto. Initiated by the Jet Propulsion Laboratory, the Grand Tour was designed to take advantage of a unique planetary alignment in the 1970s that would not take place again for another 175 years. But its enormous price tag—at around $1 billion—was not feasible to sustain. Low worked tirelessly to try to keep the Grand Tour viable, but in the end, he had to compromise in order to preserve a meaningful robotic program. The Grand Tour would eventually be replaced by Voyagers 1 and 2, launched within sixteen days of each other in 1977, to explore Jupiter, Saturn, Uranus, and Neptune. A modified version of the Mariner probes, the Voyagers had a smaller set of scientific instruments than those planned for the Grand Tour; still, they were capable of examining the outer planets in unprecedented detail.

At about this time, the Nixon administration began to request that NASA look at ways it could assist other federal agencies in solving some of the nation's biggest social ills—everything from fighting inner-city poverty to pollution to utilizing advanced technology in other federal agencies. Low was very skeptical of this scope creep. NASA's budgets were already stretched. It was a challenge to simply try to balance existing priorities, let alone layer on new ones for which it did not have particular expertise. "The domestic problems, including those of the physical environment, are largely social, economic, and political," he told a gathering of the American Institute of Aeronautics and Astronautics in November 1970. "While the need for new technology, research, and development does exist to solve domestic problems, the space agency is not the driving force to obtain a solution. The solutions to these problems must come from sociologists, economists, and politicians—not engineers and scientists. We should not think or encourage others to believe that NASA can solve all of our domestic problems."[26] Low also believed very strongly that NASA "should not take on new jobs when we aren't even doing our existing jobs in space and aeronautics as well as we should." If NASA were to get involved with anything, he argued, it should be with the advancement and application of technology in the broader economy, an activity NASA was already aggressively doing with its large research and development budgets.

"A good number of folks involved in the Nixon White House were anti-Apollo," Bill Anders explained.[27] Anders was now working in the White House as executive secretary for the National Aeronautics and Space Council, which reported to the president, vice president, and cabinet-level members on aeronautics and space policy issues. In this role, Anders would become a vital conduit and ally for Low within the Nixon administration. Like others in the administration, Anders felt that Apollo had achieved its goal and was undercutting not only robotic space programs but also aeronautics—the original mission of the NACA. Now, aeronautics made up less than 5 percent of NASA's budget—cannibalized to support Apollo. "Once we landed on the moon, we'd basically done the mission that Kennedy set out," he said, underscoring the need to pull back from the crash course to beat the Russians to the moon. "My projection was that the public support would fade rapidly, which it did. Like the Indy 500, one wondered how many times you have to drive around the track to prove to everybody that you could do it. Sooner or later, the space program had to come back down to Earth."[28]

Anders leveraged his close relationship with budget director Casper (Cap) Weinberger to help Low reestablish NASA's priorities within an acceptable budget framework for Nixon. "Cap was a closet space fan," Anders observed. "I got invited to a lot of closed-door, high-level budget reviews, where others were voting basically against Apollo or against things like the Viking spacecraft, which were taking up a good chunk of NASA's money. Cap would ask me what I thought. Even though of the six people there, five were against it, I would say, 'Well, I think we ought to do it.' And Cap would say, 'Okay. We're going to do it.'"[29]

Over the following years, Low and Anders would forge a tight, inside relationship between NASA and the White House, similar to his key relationships with people like Tiger Teague and Clinton Anderson in Congress, who helped support and keep NASA programs alive. It was an advantage that people like Paine did not have. Anders became Low's eyes and ears within the administration during this difficult time. "George was organized, and he did his homework. He knew how to get along with people, and I basically saw him in the Space Council environment," Anders explained. "I was sort of considered by the White House as a bit of a spy for NASA, and indeed I was. I could give George honest feedback about what was going on in the White House vis-à-vis space policy."[30] Low often cautioned in his "EYES ONLY" and "SENSITIVE" memos to the administrator, when sharing information he gleaned from Anders, "Once again, it is important that we protect Bill by not letting on we know any of the details of the budgetary considerations."[31] It was an effective process. Still, they could do only so much. "The Budget Bureau, Nixon's staff—they didn't give a shit about space," Anders bluntly said. As far as continuing human spaceflight with the shuttle, "all they really cared about were votes in Southern California," where a majority of aerospace manufacturing jobs were located.[32]

Low remained cognizant of this attitude throughout his time in Washington. "What we do in space must be justified on its own merits," he wrote. "This country is beyond the days of reacting to major Soviet events in space by rapidly increasing spending on our own programs."[33] Benefits and capabilities (read, "economic spin-offs") were the key sales terms in the space business of the day.

"My strategy was to save as much of the program as we could possibly carry through," Low said. "But if we must cut, let's cut back on old-fashioned things and preserve the new things. We cut back on those things we were already doing in 1970 with 1960 hardware to preserve the shuttle. In fact, funding for the shuttle was almost taken out in the fall of 1970."[34] He had to make a direct, last-minute appeal to George Schultz in the budget office, while Schultz conferred with the president to keep the shuttle studies funded. "And we kept things like the Viking mission. We were very much opposed to canceling that program. To keep it on its 1973 schedule, in those early 1970s budgets, we would have had to pour more and more money into it. So we pushed it to

1975."[35] In the end, the Voyager missions were pushed to 1977. It was a game of squeezing the balloon on the budget, but it kept both Viking and Voyager viable in the era of the shuttle ramp-up.

"When I got to Washington, the guy that really knew the shuttle was George Low," Gerry Griffin said about his time in Washington working for Low as head of NASA Legislative Affairs. Low had requested Griffin from Houston after the closeout of Apollo to help him engage a less-than-enthusiastic Congress to get them to maintain shuttle support after it was embraced by the White House. He liked to bring people to Washington who he thought had great potential for long-term careers at NASA; he felt they would benefit from the experience of doing a rotation through NASA Headquarters. Griffin's perspective from this period provides insight into Low's approach to his role as chief shuttle lobbyist. "I was thirty-eight years old, an aeronautical engineer, and I didn't know anything about the Congress. But we were trying to sell the shuttle at that point and keep it sold. So George and I did a lot of calling on the Hill. Just the two of us. We would go up and talk to the congressmen and women, and he would explain what we were doing with the shuttle, why we needed its capabilities, and why we needed their support."[36]

"We have started a series of Congressional visits," Low wrote in his personal notes about these initial efforts working the Hill. "The primary purpose of the visits is to make sure that, if asked, Congressional leaders will respond to the White House that support for the NASA program is clearly available. However, Congress is now deeply involved in so many other things, that most members would just as soon not hear about NASA." He wryly noted that a Republican senator from Colorado expressed to him the mood perfectly: "The best thing that NASA can do now is to go away and not rock the boat."[37] Anders mentions that in one budget meeting, a senior White House budget official saw Anders's model of the shuttle. "What is that?" he asked. "The space shuttle," Anders replied. "Quick—step on it and kill it before it multiplies," the man told him.[38] It was a tough sell throughout Washington.

But Low refused to cede the conversation to silence. In a memo to all associate and assistant administrators, he challenged all of them to improve their communication with Congress.

As the new session of Congress begins, I want each of you to review your system for furnishing information in response to Congressional inquiries. Over the past year, both the quality and the timelines of response to inquiries and requests for material for hearing records have left much to be desired, and they have led to repeated formal and informal complaints from the Hill. The aggravation of slow and inadequate response to requests can needlessly dilute Congressional support for NASA's budget proposals. . . . As to the quality of response, the problem is equally serious. . . . There are too many instances in which drafts furnished by offices responsible for a program or functional area need extensive revision to reflect basic facts accurately, to supply missing but germane facts, or to make clear the policy we are following.[39]

Low's message was clear, and the sense of urgency unmistakable. It was no longer a time for NASA to rest on its moon-landing laurels; the time for action was now. Otherwise, someone might step on them.

Admittedly, the negativity from the Hill could be exhausting, Griffin conceded. Low knew that most of the power and problems with setting NASA's goals resided in the White House. But it was Congress that held tight rein over the purse strings. Responding to attacks was an extremely frustrating task, but Low never let himself get down or negative about it, even in the face of strong, emotional opposition. "At times, he could get frustrated, but George wouldn't let it get to him," Griffin said, describing Low's approach to dealing with Congress. "He had control of his thoughts and emotions. He was tireless. He kept on going. His primary position was to do all he could do, to never let up. And we were fighting, really fighting, for attention in Congress. We couldn't get the Congress on a constant track. There were a few members from the districts in the states that had aerospace businesses that were always supportive, but it was hard to break into the minds of the people that represented folks that didn't have an aerospace component in their economy."[40]

Low's strategy, according to Griffin, was never to be confrontational. That was not his style. Rather, he sought to educate and let people make up their own minds. He would point out the real reason for the space program—to improve the nation's future by improving our technological and industrial base. He'd make his case earnestly. Then, he'd let the chips fall on the side of logic. "They listened to George," Griffin said. "He'd never go into a hard sale pitch. He didn't say, 'You've got to do this!' He simply said, 'Let me tell you why I think this is important that we do this.' They listened to him."[41] He was likeable. He might not have turned all their opinions in his favor, but he also wasn't thrown out of any offices, either. Like developing a spacecraft, developing suitable budgets took iterative steps, too.

During this period, Low was keenly aware of the need for speed in judiciously rightsizing the agency as its budget allocations rapidly declined. The enormous buildup for Apollo created bloat across the entire NASA supply chain and internal managerial and operational organization. Since Apollo's days were numbered and the congressional and White House appetite hinted more and more toward a low Earth-orbit future, the center budget situation demanded action. During his tenure in Washington, Low would become preoccupied by NASA workforce issues. His biggest concern from his first days in office involved attracting and retaining the young, dynamic talent that would be needed to keep the space program relevant and innovative, while at the same time transitioning the institutional base steeped in Apollo to the new priorities of the shuttle.

Reductions-in-force were a necessary evil. Low tried to minimize the impact of the reductions-in-force on NASA's critical engineering and technical foundation. In the field centers, where most of the important research and development work took place, "insofar as possible, reductions should be assigned to construction management employees, and others who are not directly involved in NASA's in-line missions," he advised his direct reports. "The reduction in terms of scientists and engineers should be minimized, especially, the young scientists and engineers, if we can help it."[42] It wasn't just that the younger staff were cheaper; with inflation, coming out of school, they were more expensive than their elders were at that time in their careers. No. It was that technology was changing so rapidly that he wanted to keep cultivating the latest innovative ideas to keep moving forward, rather than falling further behind. He would spend a considerable amount of time traveling to the field centers, interviewing and sampling the thoughts and ideas of the young engineers and scientists "to determine how best to continue their training and development of their careers." It was a concerted effort "to identify the future leaders of NASA." He viewed the young, technical minds "as superior to what Low and his contemporaries" offered NASA when they were at this stage of their careers.

John Conway was one such engineer that Low visited during his center tours. At the time, Conway was a twenty-something engineer working in the centralized computer facilities at Kennedy Space Center. "One of my bosses came to me one day and said, 'George Low is making a visit to KSC, and you've been selected to visit with him,'" he recalled. "He was meeting with half-a-dozen people like me. Instead of meeting in a conference room, he came to my cubicle. I thought, 'Are you kidding me? A guy like George Low, coming to my cubicle?' But it was true. To me, he was a great man. He managed all of NASA, internally. It was like one of the gods coming down from the heavens to meet with you." When Low arrived, he was the "most gracious, courteous, and friendly person you could ever meet." They discussed Conway's background and education, where he grew up, and how he viewed his work. "He was just very interested in me and what I was currently working on, what I thought," he said. Low told him, "I lay awake at night worrying about the top 10 percent of NASA, as well as young engineers like you. I worry about whether that top 10 percent is going to become 11 percent or fall to 9, and I worry about how they are working to inspire and motivate young engineers like you to join their ranks." Conway would spend the rest of his career at NASA, and he would often reflect on this meeting with Low. "When you look back over your life and career, there are people who intersect with you in a way that you never forget and literally become a part of your DNA. It was one of those moments for me—a challenge to become one of those 10 percent. He constantly worked that issue, trying to get more top-notch people into NASA at a time when we faced a very large challenge."[43]

In reviewing overall staffing levels, Low directed the administrator for organization and management to develop a plan to identify NASA's underemployed—those who, in this new, post-Apollo era, really no longer had a relevant job. His goal was to start identifying people to either counsel them into more productive work for the shuttle or to exit them from the organization as part of the overall process of reduction-in-force. This type of action had never been done before at NASA, and he opted to leverage in-house data to make strategic cuts rather than simple, blunt, across-the-board reductions. The data proved him correct in his approach. "As I suspected, we have very little management information that would permit us to really understand where, how, and what our workload is," he wrote in his notes after getting some initial data feedback on his initiative. "However, the little information that we do have indicates some rather disturbing trends. For example, our computer programs now tell us that throughout NASA we have 600 people working on construction management, at a time, when we have almost no construction." He instructed his people to mine the data and find other areas like construction to "cut them back drastically without really affecting our technical and engineering competence."[44]

Cuts were not just implemented at the field centers. He insisted that NASA Headquarters make a complete review of staff as well. He met individually with each associate administrator, discussing with them in detail their staffing needs. He reviewed the hundreds of positions on a case-by-case basis and "was able to get a reasonably good understanding of how each organization handled its problems."[45] Based on his assessments, he took a very large cut out of the Office of Manned Spaceflight, reducing the staff by 18 percent; he took a further cut of 10 percent from the Office of Organization and Management; and then smaller, single-digit-percentage cuts came from the other associate directorates. He also took a hard line on the use of outside consultants, slashing the budget for such activities by 50 percent. He eliminated advisory-capacity consultants, especially those serving on committees that were paid simply for providing general advice and counsel. "Only those consultants or experts who are performing a direct, in-line job in the NASA organization will continue to be paid," he directed.[46] He wanted to avoid what he called "hobby shop" and "job shop" committees, which did not have "any clear leadership" or specific goals but nonetheless seemed to have a knack for creating wasteful (and expensive) projects outside NASA's main mission and scope.

Of course, Low didn't just concern himself with cutting staff. He also quickly recognized deficiencies in NASA Headquarters that could spell doom for its future. With Mueller having resigned, there was no one effectively running the Office of Manned Spaceflight. Future, long-range planning was now nonexistent, and each budget cut meant that NASA was being reactionary rather than proactive with alternative plans. And more critically, the balance was off not only in human versus robotic spaceflight but also in the important balance between aeronautics and space programs.

He hired Dale Myers from North American Rockwell to replace Mueller as head of the Office of Manned Spaceflight. Myers would be the central figure in charge of taking the shuttle from design to production, and Low had faith in his skills and ability from his interaction with him during the rebuild after the Apollo fire. "He will be a major asset to NASA," Low wrote about his decision to hire him.[47] From a management perspective, Low was determined not to be a Mueller; he would not try to micromanage the shuttle's program office and its engineering from his ivory tower seat in Washington. That responsibility would reside with Myers and his chain of command. Additionally, he turned to Neil Armstrong and Roy Johnson to head up the reinvigorated aeronautics initiatives at NASA in the Office of Aeronautic Research and Technology. Low became convinced early in his tenure that the supercritical wing being studied at Langley "is the next major step in subsonic commercial aviation, and that it may just be what is needed to keep the United States at the forefront in this area."[48] A supercritical wing is one that is designed to delay the onset of wave drag on the wing in the transonic speed range, thus increasing the plane's fuel efficiency, among other advantages. He was disappointed in Langley's business-as-usual approach to the wing—which Airbus would eventually fly in 1972 with the A300—and he wanted to agitate for more rapid commercialization in the United States. He planned to have Armstrong "move out on a project involving the airlines to get a supercritical wing flying on a commercial airliner within the shortest possible period of time."[49]

To bridge the long-range planning gap, he went to the one man he trusted who could "put some imagination back into the future plans for the Agency." Wernher von Braun had expressed to him "a strong interest in joining NASA headquarters," even before Low was sworn in as deputy administrator.[50] Once in Washington, they held a couple of planning meetings, and they discussed the possibilities. They both agreed that the STG plan could not be handled within any reasonable amount of dollars. He asked von Braun to come up with three alternate plans: (a) a baseline plan, supported by a relatively modest shuttle program with a maximum budget in the outlying years of $5 billion, with information to detail the type of payloads to be flown and how the shuttle would be fully utilized; (b) a contingency plan, should budget levels stay at their current depressed levels and decrease, where "we will not be able to do many of the things that we have said we would like to do, and should dramatically demonstrate what this means for the future of space," including canceling much of the human spaceflight program; and (c) a bold plan, which should be available in case external influences again should prompt the question, What is the boldest thing we can do in space? "The reason we were able to get a decision to go to the moon in 1961 was that we happened to have the right kind of plan available just at the time we were asked what we could do to surpass the Russians," Low explained. "In the same vein, we should *always* be prepared to move out with such a plan in the future."[51]

Von Braun, he noted, "enthusiastically accepted the position."

While NASA's current budget situation was dire, Nixon's key aides were encouraged by what they saw in Low, and they went out of their way to remain encouraging to him. He was a pragmatist. They could reason with him. While in San Clemente, California, at Nixon's Western White House in January 1971, Low visited with Henry Kissinger, the president's secretary of state and national security advisor.[52] They were discussing an upcoming trip to Moscow in which Low would negotiate what would become the historic Apollo-Soyuz Test Project in 1975. At the meeting, Kissinger privately gave Low direct assurances that the president was behind the space shuttle and NASA's current, overall effort. He just couldn't do it with big dollars at the moment due to the current economic climate. But Kissinger urged him to have patience. They had his back.

"He said that he hoped that we would be able to hold our teams together so that when more funds became available, we would once again be able to pick up with some exciting new programs," Low recalled being told.

It was a situation he had been in before, during Mercury and the Eisenhower administration.

He was an experienced player at the game.

Be prepared. Have a plan ready.

And von Braun delivered the following month after Low returned from Moscow. Low noted,

> While I was in Houston, von Braun and Frank Williams presented to me the results of their recent long-range planning activity. They have now prepared a document which presents a baseline plan of our Fiscal Year 1972 budget and its extension, plus the new starts that we expect to undertake in the rest of the decade. Later in the week, they presented the same plan to all of the Associate Administrators and others in Washington. I was quite pleased with the results of this planning activity and suggested that we should publish the plan for internal use only and that we should keep it in draft form perpetually. In this way, it gives NASA

a baseline to measure all new activities against, without tying us down to a firm and published plan.[53]

Given the focus on budgets for NASA, Low believed that the agency and the aerospace industry needed to lower the overall cost of operations to ensure that human spaceflight would remain viable going into the next century. In 1972 he undertook a very public campaign of speeches and editorial writing on the need to attack costs. He did not like that the shuttle was being sold based on cost savings, instead of capabilities and national need. He worried that the cost figures were too conservative, that NASA's estimates might be too optimistic. He was, of course, right as usual.

At its most simplistic level, attacking the high cost of human spaceflight was the approach NASA was trying to take with the shuttle; reusing the hardware could spread out the costs for development over multiple years and flights, thus lowering the overall cost for a program. As it was now, NASA kept throwing away each rocket and capsule after they were done. The key to this working, however, would be to put continuous pressure on future operating costs— those normal, regular costs associated with flying the spacecraft. This would be key. The attention would be in the details, he knew—from design, through manufacturing, and into operations.

While NASA came in close to its original budget figure for the development of the shuttle—around $5 billion—it never approached the overly optimistic (and in hindsight, naive) figure of $10 million per flight when it went into operational flight in 1981, five years after Low left NASA. These initial cost figures were based on a study by the Mathematica Corporation,[54] a think tank that indicated NASA would need to fly the shuttle for between 750 and 900 missions over the course of its usable life to achieve the cost of $10 million per mission; in the end, the shuttle flew only 134 missions among the five orbiters built that flew into space, at an average cost to fly of over $1 billion per mission.[55] "It ate NASA hollow," Anders observed about the large cost of the shuttle to operate in the 1980s and 1990s.[56]

Cost savings from the shuttle were also predicated on the consolidation of launch facility costs, which also never fully materialized after Low left the agency. "Much has been said about the shuttle's reusability," he told the National Space Club in Washington DC in February 1972. "Less well known is the consolidation of launch service, which becomes possible with the shuttle. Did you know that we have built more than eighty launch stands in this country, fifty of which are still either active or on standby, or that we are using seventeen different combinations of boosters and upper stages in our space program today? The shuttle will allow us to consolidate most of these."[57] In the end, NASA and the military never did consolidate all these operations. "Rather than reduce the cost to orbit, the shuttle increased it by a factor of ten," Anders said.[58] It was a result that Low had feared, one he continuously cautioned about. While thousands of satellites were launched into space from the United States during the thirty years the shuttle was in operation—sometimes up to as many as one hundred a year—the shuttle only delivered about a tenth of that number, about 180, including all the payloads and modules for the International Space Station (ISS).[59] Most of those satellite payload deliveries happened before the *Challenger* accident. The number reduced dramatically afterward, and payloads mostly focused on repair and retrieval of satellites, as well as building the ISS.

"If we don't do something about the high cost of doing business in space, and do it soon, our nation's space program is in deep trouble," Low told a joint symposium of the National Security Industrial Association and the Armed Forces Management Association in Washington DC. "We are on the verge of exciting, new discoveries in space science, but we cannot follow through as rapidly as we should, because we can't afford it. We see before us many important space applications, but we cannot move out as rapidly as we should, because we can't afford it. Most important of all, we may lose our hard-won worldwide leadership in space if we don't find a way to do more for our money."[60] He admonished the industry and his own people at NASA to attack costs in the design phase—design for lower cost, standardize, don't reinvent the wheel, don't overengineer, trade features for cost, set firm cost targets, and measure against those cost targets.

Gene Kranz observed, summing up the legacy of the shuttle,

> I look at the shuttle as the last hurrah of the Mercury, Gemini, and Apollo generation, It is the device that was founded in the principles that George Low and Robert Gilruth established. Unfortunately, it has not achieved the economies that were intended. But to a great extent, these economies are not being achieved principally because of political limitations that have been put on the program. At the time of the *Challenger* accident [1986], we were one of the world's premier launchers of

satellites from the shuttle. With the stroke of a pen, it was decided that we were unwilling to risk human life to deploy satellites that could be as well deployed in an unmanned fashion.[61]

"The shuttle was a much greater reach than NASA was technically capable of delivering at the time. In fact, it was an overreach," Anders explained. "It was continually underfunded. During the operational phases, NASA management just really wasn't up to the job after George left. George would have meetings and want to hear everybody's opinion; after, management changed. They became codified; they didn't want to hear about problems. That's why we lost two shuttles, and instead of getting one flight every two weeks, we got one or two a year."[62] Despite the cost problems, the shuttle "was a magnificent flying machine," Kranz said. "I love the shuttle. It is probably the most advanced technological space system that has ever been built."[63]

On 5 January 1972 Richard Nixon formally announced the space shuttle program. "I have decided today that the United States should proceed at once with the development of an entirely new type of space transportation system designed to help transform the space frontier of the 1970s into familiar territory, easily accessible for human endeavor in the 1980s and 1990s," Nixon said. "The system will center on a space vehicle that can shuttle repeatedly from Earth to orbit and back. It will revolutionize transportation into near space by routinizing it. It will take the astronomical cost out of astronautics."[64] Space shuttle legislation was introduced in Congress, and it passed—in many ways, because of a lot of the dedicated spade work done by Low and his Legislative Affairs team behind the scenes. Nixon, like Low, believed that the shuttle would open the doors to more international cooperation in space—a centerpiece of Nixon's space and foreign policy initiatives.

After the announcement of the space shuttle program, Nixon sent Low a personal note to thank him for his hard work and express his appreciation for his "contribution to the success" of getting shuttle legislation passed by a three-to-one margin in Congress.[65] Efforts by Walter Mondale, William Proxmire, and other antishuttle advocates in the Senate were defeated, and NASA's authorization bill—including critical funding for beginning the development phase of the shuttle—passed sixty-one to twenty-one. "The debate followed traditional lines with liberal Senators from the North trying to defeat the shuttle, while a coalition of conservatives and Senators from states with space installations backed the authorization measure," the *New York Times* reported. Low's efforts behind the scenes involved saving a lot more than just the shuttle vote, as the *New York Times* article also detailed: "The authorization bill also includes $540 million for the manned *Skylab* space station, and $128 million to complete the lunar exploration program, which will end in December with the last flight, *Apollo 17*."[66]

Noted space historian Roger Launius said,

> If I were to point to the most important thing that George Low ever did, it was rescuing the agency after it had crashed and burned in the Nixon White House. When Webb departed and his deputy, Tom Paine, became administrator, he had already bought into the agency's perspective. That was very well reflected in the Space Task Group report that Nixon rejected. Paine worked hard to persuade the broader presidential commission that Vice President Spiro Agnew led that this was the right direction. Paine got so blustery that he could not even get a hearing in the White House. They refused to accept a meeting with him. And when Paine left—he was told to go, and he did—Low becomes the acting administrator. It was his task to repair the damage, which was very real.[67]

***

Peter Flanigan called Low on a Saturday morning to inform him that Nixon had selected Dr. James Fletcher as the new NASA administrator. Low immediately phoned Fletcher to offer his congratulations. They met the following Monday at the White House at a dinner to honor the returning *Apollo 14* astronauts. "My first impression of Fletcher is excellent," he wrote. "I believe that he will be very good for NASA."[68] With Fletcher not planning to start his new job until May, Low continued on in his role as acting administrator. Between the confirmation hearings in March and his start date in May, Fletcher and Low held regular planning and update meetings. "He would like for me to generally handle the internal NASA management, while he will take on the external problems," Low wrote following one such meeting. He indicated that Fletcher intended to give him a wide berth in his management of NASA and only wanted Low to bring to his attention "those areas that he should get involved in."[69]

"Fletcher came into NASA negative on manned spaceflight," Low recalled about their initial conversations. "He was selected by the people close to Nixon for being the kind of person that would support an unmanned space program. But Jim and I hit it off very quickly and easily. We talked a lot, and we went off and discussed things. I never pushed him on the shuttle."[70] Fletcher came over to the idea of supporting the shuttle on his own. Employing the soft sell, rather than the hard sell that he used with Congress, Low knew that the budgets Fletcher would have to live with for the next five years or more would be dictated by whether or not he embraced the shuttle. Fletcher needed to want to support it and not feel like it was something forced on him. Low's strategy worked. Fletcher came in not in support of it, but he very quickly turned himself around and became a big supporter of both the shuttle and manned spaceflight in general. They worked as a close team for the next five years, implementing the future space program blueprint that Low had helped forge during the last almost two years. The role Low now served for Fletcher was the sweet spot of Low's career, the same kind of role he served for Silverstein, Gilruth, and Paine.

In early 1972 Low received a handwritten note from Marsha Crandall, a young high school student from Canton, Ohio. She wanted to have some general information on "what it is you do all day" and on what, if anything, he still had to do with Apollo. Because of his many appearances in the press, Low often received such letters. Amazingly, he would always find time to answer. He took NASA's public affairs responsibilities as seriously as he did its engineering. His response to Miss Crandall gives a sense of how his typical day progressed during this period of his career.

"My day starts at 6:00 a.m., with a one-mile run before breakfast," he wrote.[71] "I am then on my way to the office by 7:00 a.m., and in the office by 7:20. The early morning time at the office is spent going over my mail, dictating, and getting some phone calls out of the way before the day's meetings begin." In describing the nature of the job, he said he had the responsibility for overall management of NASA and seeing to it that "all of NASA's programs approved by the President and the Congress are carried out. Therefore, my days are spent mostly in policy meetings, briefings and reviews of programs, decision meetings, preparation for hearings before Congressional committees, budget meetings, and planning for future programs. Some time is also required away from the office on inspection trips to our various centers, speaking engagements, and space-related functions."

He noted that he liked to use the end of the day to meet with his assistant and secretary to discuss the day's mail and messages and to map out the following day. While he tried to get home each day by 7:00 p.m., he said that wasn't always possible. And of course, there was often homework when he got home. Memos to read and answer; always, always memos.

As for Apollo, while he had general responsibility for it as a program under him as deputy administrator, it was no longer as direct as it once was when he was the program manager. You can almost sense the lament in that short answer, following the long list of administrative office duties that filled his day.

Another major departmental change Low focused on involved NASA Public Affairs. Julian Scheer, a veteran newspaper man, had been hired to head up NASA Public Affairs during Jim Webb's administration. Julian was a fierce proponent of press freedom and had done many great things during his tenure. He did not believe in puffery or any hint of marketing; he believed his office should take a newspaper approach and just deliver facts, without editorial bias. In addition, he did not think that NASA—or anyone else—should provide the astronauts with information about what they should say. Ever.

While Low generally approved of Julian's approach, he was becoming more disturbed by the lack of information that the key spokespeople of the agency—including the astronauts—had at their disposal to help them form their own opinions. Speaking their minds freely was one thing; operating blindly, without the facts or backup material on key agency initiatives was another matter. Julian refused to help, and he did not think that providing such information was his department's role. Additionally, Low observed that there was no long-range planning, no attempt to get anyone on the same message as to what NASA's purpose and goals were, and certainly no attempt to serve the NASA staff as equally as the press. "I had an eye-opener about preparing things like a briefing book in the Headquarters organization," Low noted early in his tenure, when he needed to pull information together for a French head of state who was visiting NASA. "There just are not any materials available for such a book, nor are the people really geared toward doing it. Apparently, nobody asked for such a book for many years."[72]

During the *Apollo 14* flight, in early 1971, Low spent time with the astronauts in Houston. He wanted to discuss with them NASA's future and the challenges they were having with the budget and the drop off in public interest. Low was surprised by the lack of knowledge the astronauts had with NASA's current plans. "They were concerned more with how NASA could better sell itself. It turned out they were very ill-informed about current programs and current NASA policies. They knew very little about the Space Shuttle and Houston's role in the Shuttle. They knew almost nothing about the other parts of NASA. This was quite shocking, in that the astronauts, as a group, probably get more exposure to the public than anybody else in NASA." He noted that it was clear that "each and every one of them would like to help in some constructive way."[73]

Of course, this would not be the first time Low had heard this criticism. In the fall of 1970 Tom Stafford visited Low and his family for dinner. Sometime during the evening, they discussed "at length how we can best 'sell' the space program." Stafford mentioned that the astronauts could be better utilized if they had a more planned and coordinated calendar of events for the year. "I discussed all of this with Julian Scheer," Low noted. "He immediately became defensive, and he told me that the program that we're now conducting is the best that can possibly be run."[74]

The feedback from the astronauts and Scheer's resistance to even consider developing a messaging platform seemed to be the last straw for Low. After much consideration, he decided that NASA Public Affairs needed to have a more proactive, managed approach to its activities. Julian Scheer disagreed—strongly. He was then asked to resign. Scheer left on 31 March 1971. While Scheer would go on to state that he believed he was fired because of political pressure from Olin Teague—Scheer refused to make special accommodations for some of Teague's associates at Apollo launches and even hinted that some in the White House had it out for him—others in the public affairs department were convinced that it had to do with "pressure from the astronauts."[75]

The truth was that NASA had to change to operate in a new environment, and Scheer did not want to change with it. Low wrote,

> There is a real lack in public affairs in that we do not have any kind of a research service or reference service organization. It is just impossible to dig out all of the facts that must be, and should be, at hand if we are to do a good job of telling the NASA story to the public. As far as

outsiders are concerned, both in newspapers, television, as well as the wire services and magazines, it became very clear that they don't really understand what NASA has to offer and how important NASA is to the country. This, in itself, proves that we have not done a good job in telling our story. As far as their reaction to NASA's services is concerned, this is very spotty. There are those that feel NASA has done a terrible job. My analysis of this situation is that those who were personal friends of Julian Scheer's and could, therefore, call him directly and get him to act for them, were very satisfied. However, those who did not have this personal contact or those who were disliked by Julian, were just completely unable to get anything out of NASA.[76]

It was time for a change.

Low wanted NASA to become more aggressive in its approach to public affairs. He commissioned Paul Sawyer, a lawyer who represented the astronauts in their outside commercial dealings with publications such as *Time/Life*, to conduct a study of all NASA's public affairs activities. Some thought it odd he'd go to the astronauts' personal lawyer, but it seemed like a natural source for Low. Sawyer dealt with most of the publications that were syndicating the astronauts' stories—he had an in with their editorial boards. It was to be a comprehensive report. The report solicited input from the current NASA Public Affairs staff, outside media contacts, the astronauts, and select media representatives from the contractor organizations. Sawyer's analysis fell in line with Low's thinking, but it also challenged the assumption that the astronauts should be NASA's leading spokespeople: "My survey has indicated that even as highly sophisticated a group as the executive media personnel are shockingly unaware of rudimentary facts concerning the benefits and applications of the space program. . . . It is the opinion of the substantial majority of the media and public relations people who I interviewed that there has been a public relations failure by NASA to explain the space program, its objectives, its present and projected benefits, and its relationship to the problems and issues in the public at large."[77] Sawyer also took Low to meet with the managing editor of *Life* magazine. It was an eye-opening experience for Low. They discussed whether or not it made sense for the astronauts to be the main agency spokespeople—if yes, then human flight (and its missions and expenses) would always dominate NASA news in the media at the expense of the important robotic, scientific programs. There needed to be a shift in focus, especially in light of the upcoming hiatus in human flight in the transition from Apollo to the shuttle. Also, NASA Public Affairs needed to become more proactive and responsive. "The main thing I got out of the meeting is that NASA, as seen through the eyes of a customer of the Public Affairs office, is a completely non-responsive outfit," Low said. "We don't answer letters; we don't comply with requests; we don't even give a 'no' answer; we just let things sit. There is no other government agency that is as non-responsive as we are. When we do talk to people, we are polite, but nothing happens."[78]

When Fletcher joined, he agreed with Low. In memo exchanges between them, they both discussed the need to have people in NASA Public Affairs who are not just news people but rather "marketers, people who know what the customer, i.e, a media representative, wants and are able to convince him that NASA can pull together whatever is needed for his story."[79] Working with the White House, Paul Sawyer, and their contractors and media contacts, NASA eventually settled on John Donnelly, a "young, but bright" PR official from a contractor. (Low considered other, more seasoned executives from advertising and PR agencies; however, NASA's inability to match private-sector salary ranges impeded this effort.) Instead of reporting into a line organization, Donnelly initially reported to Low, who worked with him on long-range planning and pushed for the office to be molded in a new light. NASA Public Affairs was to be not just a conduit of facts but also an office that actively sought to market NASA's story.

In a post-Apollo world, his more aggressive approach to public affairs appealed to people like Doug Ward, the public affairs officer who worked with Low in Houston. "He was a genius at perceiving public and congressional perceptions, and he knew what kinds of information needed to be developed to deal with that," Ward recalled. "When he went to Washington, he elevated George English, a young engineer, out of the mechanical division to assist him. He was a bright, very articulate person. And when Low would encounter something that he thought needed clarification or where there was a public relations issue, he'd send English off to develop the background facts to support it. And then Low would usually articulate that response in a speech."[80]

Ward specifically remembers Low working with English to provide statistics to defend spending money on human spaceflight.

> Critics were constantly bringing up that we shouldn't be spending all this money on manned spaceflight. Low developed a set of background facts that showed that unmanned programs were not adversely affected by the manned program; in fact, as manned program funding increased, so did the funding for unmanned. Once he developed something like that and articulated it, it became kind of the template that the public affairs world throughout NASA would pick up on and use. I know of no other top manager who was as adept or as sensitive to those kinds of issues as Low. Once he left NASA, we really lost the ability, I think, to respond to those kinds of issues in that forceful, direct way.[81]

Of course, personnel issues also dominated his time in the 1970s with the constant layoffs across NASA's entire employee base. On a day-to-day basis, he had his hands full with such activities as meetings, updates, and decisions. Labor unions sued; downsized contractors sued; congressional members fought for reinstatement of people and facilities in their districts; and friction was starting to form within the agency between the "Apollo people" and the "shuttle people." But no issue raised itself to the level of congressional scrutiny and bad press more than the issue of NASA's lack of women and minority representation both in field center technical positions and in the very visible astronaut corps. The reductions-in-force had exacerbated the issue, as the representation was already low in the technical areas. With non-technical areas taking the majority of the layoffs, NASA's equal-employment numbers were abysmal.

NASA had an Equal Employment Opportunity (EEO) effort underway at the time, even before the passing of the Equal Employment Act (EEA) of 1972. But NASA's efforts were lagging behind other federal agencies. After all, it did not even change the *Manned* Spacecraft Center to the Johnson Spacecraft Center until early 1973. In charge of NASA's EEO was Dudley G. McConnell. Low had a reputation for encouraging women and minorities who did not have an advanced education to go back to school to pursue their degrees. NASA was a technical organization, he would counsel, and to get ahead, one needed an advanced degree—regardless of race or gender. McConnell was one such person he took under his wing. He hired McConnell, a young, African American mechanical engineer at Lewis in Cleveland, in 1958. According to McConnell, Low mentored him and encouraged him to continue his studies and get his doctorate in aerospace physics. He also brought him to Washington as his career progressed to increasing levels of responsibility.[82]

After the passage of the Equal Employment Act, Low put McConnell in charge of NASA's EEO efforts. NASA Headquarters had improved on its hiring of women and minorities—bringing their percentage of the workforce up to the double digits, in raw percentage terms. Still, most of those jobs were in relatively low-level clerical functions. The field centers, where most of NASA's higher-paying technical jobs resided, were performing the worst. Most were in the single digits, in percentage terms, and it brought the agency's overall average down to the lowest level of all federal agencies.

On McConnell's staff was famed African American EEO and human resources expert Ruth Bates Harris. Fletcher and Low had already inadvertently bungled her appointment to the position, generating confusion over her title and role within the organization. She was first to be hired as the head of EEO but then, shortly after joining, was made a deputy assistant to McConnell in January 1972. Additionally, Harris and McConnell did not see eye to eye. (In congressional hearings investigating the situation, for example, Harris reported that McConnell used to call meetings to order by ringing a bell, an act that Harris found deeply offensive and demeaning.) Despite several years of effort, Harris believed that NASA was seriously lagging in its programs and commitments. On her own, she worked with several EEO colleagues and authored a frank, blistering forty-page report in late 1973 about NASA's lack of progress. She recommended the replacement of McConnell and rightfully blamed the rapid march toward the moon for NASA's ignoring the representative imbalance in its labor pool.

"For many years the program of getting a man on the moon took precedence, and top priority over everything else," she wrote.[83] "But looking at NASA plans for the future, we find that an entire generation of people will have been cheated from witnessing such experience as seeking a woman or a minority in space." As for women in space, Harris pointed out that there have been three: Arabella and Anita, two spiders used in an experiment, and Miss Baker, a monkey that flew before Alan Shepard. It was a scathing, eye-opening report—and one she delivered personally to Administrator Fletcher.

Fletcher's response? He fired her. The action became a media firestorm, with editorials claiming that NASA had "killed the messenger."[84]

Senator Proxmire used the situation to stage a scathing congressional hearing, airing out NASA's poor record on the issue. He called NASA's efforts "a total failure." It was a wake-up call. Like the *Apollo 1* fire, Low knew that blame did not rest with just one or two people; NASA's poor record in aggregate was an institutional problem that needed to be corrected. Not turning his back on the problem or seeking to deflect blame, Low leaned into it. It weighed heavily on his mind. "The facts are, we did not do an adequate job in previous years," he remarked about the issue. "Government wide minority employment was at 20 percent; at NASA, we were at just 5 percent. As an Agency, we had a lot of catching up to do."[85]

A first step was to heal the open wound caused by the firing of Harris. Low worked with Harris, Fletcher, and Harris's lawyers, and NASA hired her back in 1974. "I am quite enthusiastic," she said at the time.[86] "This is the kind of assignment I have wanted to be involved in. I'd like to forget the past, and think about the wonderful future we have." To further support the EEO effort, Low and Fletcher also hired Harriet Jenkins, another accomplished African American woman in the field of equal employment, to work with McConnell. While McConnell had concerns that Jenkins would outdo him in performance—that she was smarter than him—Low was adamant that she would be a great resource for him. "He needs the help," Low bluntly wrote in his notes.[87] Jenkins was hired in as an assistant administrator, rather than a deputy assistant administrator, as Harris had been. "Now that's progress," Harris said of the hire and title.[88]

As a team, they worked to tackle the very real problem—not just in staff positions but also in preparing the agency to recruit women and minority astronauts as well. "I worked most closely with George Low," Jenkins said. "We had an excellent relationship. We launched excellent initiatives, and they were very effective. But it was the result of teamwork. The policies were being directed by the leadership of the agency. We were collaborating and planning together, so George and I got along well in that kind of process."[89] The problems, of course, were very real. Women and minorities were underrepresented in technical fields in general, let alone at NASA. The hiring pool was thin, and private industry—with better pay scales—provided tough competition for talent. "NASA had underrepresentation of minorities, women, and disabled individuals, especially in the scientific and engineering fields; that was the commonality," Jenkins explained. "It was also challenging at the time because the very disciplines that NASA needed were not overpopulated with minorities and women."[90]

Along with new EEO programs within NASA, the agency's focus turned to working with universities and secondary education institutions to create programs to assist and encourage women and minorities to go into technical fields. "I remember people really wanting to get the job done," Jenkins said about accepting the job to work with McConnell and Harris. "There were excellent programs that were being implemented. When I looked them over, certainly Dr. McConnell and Ruth Bates Harris had some excellent ideas that they had initiated and were in place. I felt I could help and wanted to get on with the work."[91] Jenkins and the overall EEO function finally started to gain traction not only within NASA Headquarters but also in the field centers.

It was Low's firm belief that the shuttle era, with its need for mission specialists and scientists, would offer the best opportunity for minorities and other underrepresented groups to fly in space. He was right. After Apollo, NASA's first group of astronauts was selected in 1978. Thirty-five new astronauts were chosen to train exclusively for space shuttle missions. It was NASA's most diverse set of candidates; the class included NASA's first three African American male astronauts, first six female astronauts, and the first Asian American astronaut.[92] In 1983 Sally Ride would become the first American woman in space, and Guy Bluford, the first African American male in space. Both were alumni of the group hired in 1978.

Despite his focus on day-to-day administrative duties, Low kept himself personally involved in all the Apollo lunar landings, through to its final mission in December 1972. While he wasn't involved directly in the program management, he did go to every launch. He took his seat within the glassed VIP lounge in the back of mission control. He privately complained, of course, that he disliked being merely an observer, having to simply socialize with the other visiting dignitaries and guests. Always the dirty-hands engineer, he missed being in on the action with Kraft and Slayton; he often found his way to the floor.

During the nearly fatal *Apollo 13* mission, all hands were made available in the effort to safely bring the astronauts home. Low worked with fellow engineers in mission control and also conducted a number of press conferences to discuss the situation as it unfolded. While those in mission control concerned themselves with *Apollo 13*'s consumables, he admitted in his notes to being more concerned about their use of the LM as a lifeboat.

> My main concern during the flight, after the accident, was not the availability of consumables, because this was calculated quite early, but the wellbeing of all of the Lunar Module's systems. We were flying in a condition where everything had to continue working from now on. If we had lost any major component or subsystem in the LM, of course, the mission would have ended in a disaster. And none of the LM systems had been required to work for the length of time that they were working and under the power conditions that they were working in this flight.

Thankfully, he wrote, the spacecraft performed admirably, and all the hardware "worked exceptionally well and helped lead to a successful landing."[93]

Given his experience with the fire, he focused on the postflight analysis, especially in light of the four remaining lunar landing missions. Was there something they had missed? A new lesson to be learned? As with the fire, the situation once again came down to not asking the right questions. On 15 June 1970 he examined the report by the *Apollo 13* review board, headed by Ed Cortright. In the report, "it had been conclusively determined that the basic cause for the accident was the procedure used pre-flight in de-tanking the oxygen tank." McDivitt, the Apollo program manager, did not switch out the tank after a detanking anomaly had been detected. This was because he was operating on incomplete information. He had not been informed that the tank had been dropped or jarred at the time of his decision; had he been, he might have swapped out the tank. "In reviewing the entire accident, it is apparent that a series of mistakes occurred, starting with the specifications for the thermal switches and inadequate understanding of the effect of high-pressure oxygen on the materials in the tank, insufficient knowledge by McDivitt when he decided not to change out the tank, and finally, the procedure itself which was not adequately understood. The whole situation can be summed up by the statement, 'We did not ask the right questions.'"[94]

The findings of the report were later submitted to Congress. Hearings before the Senate in Washington were held on 30 June. Like everyone else within NASA management, Low was concerned with how congressional and media response might impact the agency's image. Had they learned anything from the *Apollo 1* fire? "The results of the hearings were quite positive," Low wrote in his notes, echoing the support Congress showed N A S A, McDivitt, the astronauts, and the entire N A S A team. "There was no significant criticism of N A S A, and only praise for the way N A S A handled the *Apollo 13* situation. Newspaper and editorial comments ended very soon after these hearings and, insofar as the public and the Congress are concerned, the *Apollo 13* situation is just about forgotten."[95]

Forgotten. Interest in the space program was at an all-time low until the near tragedy of *Apollo 13*. The human drama of the event and N A S A's successful handling of it—including allowing the press into mission control to cover every moment of the situation as it unfolded—brought a momentary peak of interest again on a global scale. Afterward, however, interest began to fall, becoming "just about forgotten." Rather than a grand adventure story, Apollo was now a geology and science story—stories that are very tough for a mass audience to appreciate.

Apollo 14 represented a successful return to flight with America's first space explorer, Alan Shepard. As one of the people who told Shepard he would never fly again after being diagnosed with Ménière's disease, Low was personally very gratified to see him now lead such an important mission. "With very few exceptions, even our critics hailed the success of *Apollo 14* as a demonstration of man over machine, and a demonstration of what an explorer can do in space," Low noted, his excitement about the mission's success clear in his words.[96] "Editorial comment around the country, including the *New York Times*, was most favorable. Shepard, Roosa, and Mitchell had done an outstanding job, not only in carrying out the mission, but also in the way they described it as they went along. The trek to Cone Crater, I think, will become a classic example of how man can conquer his environment." Still, worldwide interest was not as strong as with the high drama of *Apollo 13*. N A S A, it seemed to the public, was back to successfully bringing back bags of rocks.

Of all the later Apollo missions, *Apollo 15* became Low's most admired and most disappointing—almost simultaneously—because of the hopes he had for the crew and their mission to engage the public with a positive image for N A S A. He admired the mission at the time of the flight because, he believed, it represented a watershed moment of science and public interest in the program. He held the crew in high regard, especially its commander. "I went to the Cape for the launch of *Apollo 15*, and then on to Houston after transposition and docking," Low recorded in his notes.[97] He spent the remainder of the mission in Houston, and among many other things with this flight, he believed that it would reveal "a scientific return beyond all expectations" and "a reawakening of the public interest in Apollo." Low attributed amazing accomplishments to both the skill and performance of the crew. "The scientific return stems in part from the augmented capability of both the Lunar Module and the Command and Service Module orbital experiments," he wrote. "However, it is also due in a large measure to the fantastic efforts made by the three crew members in learning the science and explaining it as they went along and to the leadership exhibited by Dave Scott." Low believed the public reaction had to do with the crew's "intense interest and expressions about the moon as they went about their business" and with the fact that they took the television camera with them as they explored the lunar surface with the lunar rover. "I attribute our feelings about the flight much more to the fact that the television camera went along with them than to the fact that they were driving all over the moon," Low observed. He was so excited by the mission, and because of "the overall attitude expressed about the moon" in the media, that he believed NASA should reinsert developing a lunar base as a long-range goal in the same sense that the STG post-Apollo plan had focused on landing on Mars. "Even though people were not able to relate to Mars and a Martian goal, I believe they would be able to relate to a lunar base goal for the 1980s," he said, especially in light of the excitement generated by *Apollo 15*. Several months later, his excitement was undiminished. He attended the *Apollo 15* lunar science results presentation by the crew and MSC lead geoscientist Paul Gast, in Houston. "The results are extremely exciting," he noted.[98] "Based on the information that is just now becoming available, Gast is beginning to define in detail the entire structure of the moon. . . . My own basic conclusion after spending a fascinating morning listening to this discussion, is that we are on the verge of understanding the moon in all of its detail."

As he had experienced after Al Shepard's first flight, as he had experienced with Neil Armstrong and Dick Gordon in South America, and as he had tried to get Julian Scheer to understand, the astronauts were NASA's best marketing tool for engagement and public awareness for human spaceflight programs. His hope was that the *Apollo 15* crew would mark a resurgence in that engagement—that is, until rumors started to percolate about commercial interests having pushed their way aboard the flight and until a number of serious allegations were made regarding unauthorized "smuggling" of mementos during the mission. It was not long before the press got wind of the stories and the mission began to draw very negative and damaging attention.

In recalling the postmission problems and their personal impact on him, Low stated, with resignation and disappointment, *"Apollo 15* hurt me deeply."[99] Shortly before the commercialization issues with *Apollo 15* came to light, problems were starting to gain attention in Congress and the press concerning items from the astronauts' personal preference kits (PPKs, also known as APKs, for "astronaut preference kits"). These were fireproof beta cloth bags in which astronauts could store personal items as souvenirs and mementos of their mission. "The situation concerning the astronauts' personal belongings carried on space flights is still not under control," Low noted in January 1972.[100] "First there was the story of the coins used for commercial purposes by the Franklin Mint to attract new members. These allegedly had been carried to the moon on *Apollo 14*. Then there was a question of the flag carried on *Apollo 14* being raffled off for fundraising purposes by the Muscular Dystrophy Association. Most recently, Aldrin told me that 'his' watch, the only watch on the surface of the moon on *Apollo 11*, had been stolen during the transfer from Houston to the Smithsonian, where he was going to provide it on loan." As a result, he noted, the Astronaut Office "issued a policy concerning the items to be included in the astronaut's preference kit, clearly stating that these were for their personal use and not to be used for any commercial or fundraising purposes."

A few months later, rumors began to swirl that the Bulova watch company had convinced Dave Scott to take a couple of unauthorized Bulovas to the moon—a wristwatch and a stopwatch. There had been an intense competition between Omega and Bulova to be the "official watch" on the moon. Omega, a Swiss-made watch, had been tested and cleared to fly. Bulova—at the time, the only American-made watch trying to get aboard a mission—had not. But that did not stop General McCormack, a Bulova board member, from trying his best to get a watch aboard a flight. "After checking out some rumors, we discovered that the *Apollo 15* astronauts carried onboard some equipment that was strictly against the rules. This is particularly surprising," Low noted, "since this came on the heels of the *Apollo 14* Franklin Mint episode, and Dave Scott had told Fletcher and me, during our visit to the Cape before the flight, that Deke had really cracked down and they were severely limited in what they could carry."[101] Low went on to detail how McCormack convinced Scott to carry a Bulova wristwatch and stopwatch "after McCormack had failed in all official approaches to NASA requesting that we substitute a Bulova watch for the Omega." He then detailed the initial indications that "the crew carried 400 first day covers to the moon" and that "the rumor was that they got $50,000 apiece for carrying them," although "Slayton claims that they received no money for them." Additionally, Low details that the crew gave one hundred of the postal covers to a German friend of theirs, "who was now selling them for $1,500 apiece." Low noted that NASA had confiscated the covers and was investigating.

Low tried to get more information about the situation via MSC management. He had Dale Myers collect and report "the facts as they were known," but "the series of facts since . . . proved to be incorrect in many details." The stories were widely inconsistent, especially about whether or not the astronauts profited. In an addendum to Low's personal note number 74, marked across the top "VERY SENSITIVE," Low details the complete cover situation in minute detail—especially the fact that he felt misled by Scott about the original intent of the crew to receive payment for the covers.[102] After an extensive investigation, NASA management—including Fletcher, Low, Kraft, Shapley, Dale Myers, Gen. A. J. Russell, and Bob Seamans—concluded that the astronauts acted "stupidly" and that they needed to be reprimanded. "I am not yet sure whether we will be able to really forgive this stupidity without getting into a great deal of trouble for NASA," Low noted.[103] Seamans remarked that the U.S. Air Force "would have a very difficult time dealing with the situation" and that the astronauts "were essentially unpromotable." Since the air force "would not be able to find suitable jobs for them," the team decided to reprimand the astronauts and remove them from flight status but keep them with NASA so that they could at least find suitable work. While a committee had decided their fate, it would be left to Low to meet with the astronauts eye to eye and issue their formal reprimands.

Irwin had already left NASA by this time, but Scott and Worden were still active. Scott would be reassigned to Dryden Flight Research Center at Edwards, and Worden would be sent to Ames Research Center. In the meantime, Low received angry letters from citizens attacking the astronauts and their "abuse" of taxpayer money. There was a lot of swearing and vitriol aimed at both the astronauts and NASA. Low would answer the letters in a calm, professional manner, defending the astronauts while also acknowledging that what they did with the stamp dealer was a terrible lapse in judgment. In one such letter on 26 July, he wrote,

> Deplorable as was the action of the *Apollo 15* crew, it did not result in personal gain for them or for any member of their families. They have been publicly reprimanded, and this fact has been entered into their personal records. Even though this may not seem to be severe enough punishment in your judgement, the future careers of these men are most adversely affected. I can assure you that no one was more distressed over this incident than I or other top NASA officials. . . . In the case of the *Apollo 15* crew, they gave in to temptation, but fortunately realized their mistake and made a clean breast of it. These were brave men; they did perform their mission well and to the glory of the United States.[104]

He hoped the situation would die down, but it would not.

After press coverage of the issue, Sen. Clinton P. Anderson, chairman of the Committee on Aeronautical and Space Sciences, decided to hold congressional hearings, despite Fletcher and Low's attempts to dissuade him.

> Senator Anderson, who apparently has been a crusader all his life (he surfaced the Teapot Dome scandal), is quite bitter about the whole astronaut first day cover business. . . . Fletcher and I visited with him . . . to try to talk him out of any kind of a hearing. . . . We made clear to him that we had no objection to holding such a hearing, but we felt that the astronauts had been punished more severely than they would have been were they not so much in the limelight, and that it was really unfair to crucify them in public [any further].[105]

In a nod to their efforts, Anderson agreed to keep the hearing closed door and in executive session. At the hearing, NASA management was raked over the coals and blamed for not doing a better job of protecting the astronauts. "The hearings themselves did not go well at all," Low noted, pointing out that the NASA inspector "made a very poor witness. He was imprecise, vague, and just did not look like a strong investigator. Fletcher and I took a severe beating on NASA's administrative procedures. Chris Kraft took a very offensive approach to the Committee, which of course, annoyed the Committee. Deke made a fairly good witness, but the Committee did not believe his story. The astronauts themselves appeared to be even more naïve than they actually were."[106] As a result of the hearings, Congress demanded that NASA develop a list of all the personal and mission items that the astronauts carried, and they required them to tighten up their administrative procedures. They tightened and reissued their rules on Apollo PPKs. In the end, NASA would crack down and strictly limit the number of personal items that could be carried to just twelve items on *Apollo 17*, the last lunar landing. "Immediately after the hearing, Fletcher and I decided that we must take early action on new, and extremely tight regulations," Low noted.[107] "The significance of the new rules which would limit the number of items carried to 12 becomes particularly apparent when one recognizes that on *Apollo 15*, the astronauts had carried nearly 2,000 personal items into space."

Press coverage was overwhelmingly negative—from poor headlines about astronauts cashing in, to antispace editorials, to editorial cartoons about the moon being for sale. Unfortunately, further investigation would reveal another negative commercialization issue with the same German stamp dealer; most of the active astronauts, including such high-profile luminaries as John Glenn and Neil Armstrong, were selling their autographs to him for five dollars apiece.[108] After further investigation, Low found that NASA did not have a specific policy against accepting payments for autographs, as it did for selling flown artifacts. Those involved in the autograph incident escaped the harsh reprimands and removal from flight status that befell the *Apollo 15* crew, except for Jack Swigert. Swigert, who famously received a reprieve for forgetting to file his income tax while on his way to the moon on *Apollo 13*, was caught up in an alleged tax dodge with a charity for his signature payments. His initial story to Low rapidly unraveled, and it was clear that he had misled management and more directly violated NASA's code of conduct than the others. It cost him his flight status. He was removed from consideration for the Apollo-Soyuz Test Project (ASTP). "We are still finding new astronaut [commercialization] problems each time we turn over another rock," Low lamented in his records, clearly being both annoyed by the distraction and let down personally by the astronauts.[109] In the end, Fletcher issued rules and regulations around autographs, similar to the rules around PPKs and NASA official codes of conduct, noting that "nothing of value should be requested or accepted in exchange for an autograph" while an astronaut was in the employ of NASA.[110]

The entire situation could not have come at a worse time for Low and NASA. They were still battling the opponents of Apollo within the Nixon White House, as well as the opponents of expensive human spaceflight within Congress. Funding for the shuttle and NASA's future programs continued to get hammered by Proxmire and Mondale in the Senate, and now Clinton P. Anderson, a longtime supporter, was "beyond irritated" with the astronauts' "shenanigans." Swigert took a leave from NASA to work for the House of Representatives as executive director of the Committee on Science and Astronautics. According to Low's notes, it was believed that he bent the ear of such NASA supporters as Olin Teague over the poor treatment of the astronauts during the scandals. It was not helpful.

The astronauts, for their part, felt that NASA management was letting them down. Peter Sawyer, the astronauts' personal-contracts lawyer, approached Low with the astronauts' collective concern. In a private meeting in Low's office, Sawyer stated that the astronauts whom he had talked to felt that NASA had not "demonstrated leadership" in "supporting their cause" and that NASA, in reporting the facts about the commercialization issues, was not being sufficiently positive about the astronauts and was "tearing them down while reporting the facts."[111] In addition, Sawyer said the astronauts "felt that NASA should start an active campaign to rebuild the astronauts' image." Low was incredulous. But he remained calm, as he always did. He simply smiled and said, "Only the astronauts themselves can rebuild their image. They need to start realizing this." It was not NASA entering into these deals; it was the astronauts. If they were sophisticated enough to get Sawyer involved in their "official" commercial dealings, they should be just as eager to get him involved in the "unofficial" ones for guidance.

Still, Low said he would consider their opinion on any further announcements related to the issue. He also stated that he would personally fly down to Houston and meet with the astronauts, which he did on 27 September 1972. "The meeting went well," Low wrote, after a long and detailed meeting in which he not only discussed the commercialization problems but also engaged them in intense talk about the future of NASA.[112] It was about as understated as he possibly could be in his personal notes. He noted the details of the meeting in his SENSITIVE FILE, going chapter and verse into the repercussions

of the astronauts' lapse in judgment. There would be no more commercialization to haunt them, or NASA, going forward. "The *Apollo 15* shenanigans
have immeasurably damaged the astronauts' image," Paul Sawyer told Robert
Sherrod in a private interview two months after Low's meeting with the astronauts in Houston. "That thing caused enormous problems, terrible repercussions. The heroes have been damaged; NASA has been damaged."[113]

As a result, Low gave up on reintroducing a lunar base as a NASA long-
range goal. With the Senate space committee still angry, all their effort would
now be on repairing the damage and maintaining existing funding levels—
especially for the shuttle.

The moon was now a destination in the past.

For Low it was time to move forward on burnishing NASA's image. Taxpayers
expect a return on their investments in programs like NASA. It was time, he
believed, to move out beyond the uninspired attempt at just promoting spin-
offs. A Teflon pan was not a reason to go to Mars. Similar to his spaceflight
cost-reduction and productivity-improvement campaign, Low for a long time
believed that NASA did not emphasize the benefits of spaceflight enough to
the general public. While Congress and Fletcher both wanted more emphasis
on technology and product spin-offs, Low did not think such an effort made
the most effective argument for the nation to spend precious resources on the
space program. In his mind, you could not sell a product by its by-products.
Space, he firmly believed, had to be sold to the public based on the benefits
it provided the nation. "It is a departure from my usual cost-savings speeches
that I have given for the past two years," he said, but it was important that people start to recognize the return on their investment in the space program.[114]

In 1973 he went about a concerted PR campaign of getting his message
out via a series of interviews with people he considered to be space buffs
in the media, who would present his view of the program to the public in
a positive light. For his speeches, he used the backdrop of the successful
1973–74 *Skylab* missions, in which the sun was studied in unprecedented
detail and the Earth Resources Experiment Package (EREP) viewed the
earth with sensors that recorded data of our home planet for the first time
in the visible, infrared, and microwave spectral regions. He emphasized the
ongoing research and application of recently hard-fought advancements in
spaceflight technology to perform critical onboard experiments that not

only would benefit future space exploration but were also helping people back home on Earth.

Of course, he also admitted that most of the benefits from the space program are broad, long-tail benefits—benefits that stretch across society and impact in a myriad of ways for generations. "When discussing the benefits of the space program, it is helpful to distinguish generally between the different types of benefits that are realized," he told Bob Thomas of *Today*. "First, there are the broad benefits of space science and exploration. These are difficult to define, but they are nevertheless real. Today we are making practical use of scientific research conducted thirty to fifty years ago. In the same way, we can be sure that thirty to fifty years from now, our children will be making practical use of the results of the science and exploration we are doing today, even if we cannot say exactly how this will come about."[115]

Naturally, he also spoke of the current benefits, especially with direct application to the advancement of aeronautical and space technology and systems: "The improvement in civil and military aviation, improvements in communications and weather forecasting from the use of satellites, and the benefits in many fields that will be obtained through the use of earth resources satellites" were already improving people's lives.[116]

The third leg of his platform avoided the usual NASA spin-off pitch. Low was more interested in the important role that NASA's industrial investments realized for future productivity gains, pointing out that for every dollar invested in research and development in the space program, the economy realizes more than seven to eight dollars in economic benefit over the following decade. Low, like many other people in this decade, started to become concerned with the threat that foreign competition made to America's industrial base. Japan and other low-cost producing nations were starting to take on more market share, and U.S. manufacturers were struggling to respond. Quality was slipping. Aggregate U.S. productivity was steadily falling after the first few war-boom decades. "The increasing competition in our economy means we must continually increase our productivity, and the only way to do this is through advancing our technology. This is another important aspect of the economic benefits that flow from our programs in aeronautics and space to the broad economy."[117]

Low took his speech on the road to aeronautical industry conferences. At a dinner for the American Institute of Aeronautics and Astronautics in June

1973, his keynote speech focused on the key economic benefits of space exploration. "When the Wright brothers first flew in 1903, many people asked, 'Why?'" he told the crowd. "They said the same thing again when Charles Lindbergh flew nonstop across the Atlantic in 1927. People now ask about the moon landings in the same, skeptical tone, 'Why?'" He named some of the recent practical applications in science and medicine that were currently providing major benefits, before he centered on the field of communications. "Breakthroughs are already being made. For example, fifteen years ago, no one would have predicted color television broadcasts from China would have been possible." He discussed the use of satellites for voice communication, over the antiquated use of undersea cable systems. He discussed solar-energy applications, pointing out that NASA was now exploring "ways of collecting the sun's energy" for practical use on Earth. Satellites, he said, while showing the audience a picture from space of sewage spills off Long Island, can also spot acid slides and sewage dumping and thus can provide officials with the tools to respond and combat these environmental problems. "Thousands of other lives could be saved by long-range weather forecasts that would warn of oncoming hurricanes, tornadoes, earthquakes, and other natural disasters," he said.[118]

In medicine, he alluded to "a home-based medical system" that only now— some fifty years later—is coming into realization with the advent of personal biometric monitors (e.g., Fitbit) and smartphone health apps. In 1973, though, he already foresaw such possibilities: "My predictions might be conservative," he said. "But space-age computers, sensors, and satellites will provide the means for an individual to give himself a physical examination in his home, by himself. The results of the examination will be fed into his own telemetry system, the results 'read,' and an instant diagnosis made of the individual's health. And if a prescription is needed, the system will instantly print it out." He discussed medical databases and systems that would be "linked to the home and other telemetry systems," providing instant access in the case of emergencies. He prophesized that communication satellites would not only provide help in scheduling ships and airplanes in a similar fashion, but these systems would also "keep libraries up to date" and "educate millions." The economic impact from these and many other as-of-yet-unknown space-adapted programs, he said, could yield "billions of dollars in annual benefit" for the economy.[119]

"Space applications is where the future is," he proclaimed, receiving a rousing round of applause.[120]

***

The last mission flown with Apollo hardware also marked the first major, albeit tentative, steps toward international cooperation in space. The ASTP in 1975 resulted from off-and-on efforts behind the scenes at NASA that began shortly after President Kennedy publically hinted that the moon race be a joint U.S.-Soviet venture in 1963.[121] ASTP informally signaled the end of "the great space race" competition between the world's two largest super powers, and it marked a dramatic, symbolic thawing in the cold war that had dominated geopolitics since the end of World War II. Once again, Low's involvement at a critical point in time would prove pivotal in making ASTP a reality. In addition to ASTP being the last Apollo spacecraft to fly, it would also be the last manned flight of his NASA career.

"The idea of cooperation with the Soviet Union stemmed from a general desire for NASA to work internationally," Low said.[122] President Nixon had told Low that he was interested in making the space program a truly international one and that he wanted NASA to stress international cooperation and participation for all nations.[123] The Europeans were interested in working on the United States' post-Apollo initiatives. Low was working closely with his counterparts, but he was not making much progress. Most of NASA's future plans sounded "too vague," and talks remained purely in the theoretical phase. The Soviets, however, were another matter.[124]

Low had a relatively long, if tentative, relationship with the Soviet side. During Apollo he had exchanged some telex messages with members of the Soviet Academy of Sciences, in which he congratulated Russia on their Luna robotic missions to the moon. Back in 1959, during Project Mercury, Low gave Soviet Academy of Sciences guests an unprecedented tour of the Langley Research Center in Virginia. During the visit, he showed them a detailed model of the Mercury spacecraft. They discussed the Soviets perhaps, one day, returning the favor in Russia. The scientists "were polite but non-committal" and "the hoped-for-invitation to see Soviet space-flight facilities never materialized."[125]

In *The Partnership: A History of the Apollo-Soyuz Test Project*, Edward and Linda Ezell credit Tom Paine for initiating the events that would eventually result in ASTP. He and the Soviet Academy of Sciences president Mstislav Keldysh increasingly exchanged letters in order to establish a direct line of communication to initiate some form of cooperation. When Paine resigned as administrator, Low took over negotiations on behalf of the United States. Keldysh had invited Paine to come, but Low felt it too premature.

"As soon as we realized the Soviets were interested, we put a study team together," Low explained. "We were considering what could be done with space stations and *Skylab* activities. I asked Bob Gilruth to head the team, as the most knowledgeable person in manned spaceflight. He was inventive in docking systems, and it was his people who had worked on that, so I thought he would be the best one to head the work group. The general agreement signed out of that effort talked about docking systems for future generations of spacecraft. When Bob got back, he and I discussed it, and we both agreed it would be much better if we could do something with present spacecraft."[126]

"Our initial response was 'cautious optimism,'" Low further explained. "Until the initial words were matched with deeds and hard facts and hard negotiations, we did not want to jump to any conclusions."[127] After the cautiously optimistic meetings with Gilruth's work group, Keldysh once again invited Low to Moscow in January 1971. After conferring with the White House, State Department, and others, he accepted the offer.

In preparation for his trip, he traveled to San Clemente, California, to meet with Henry Kissinger. Speaking on behalf of Nixon, Kissinger gave Low wide latitude to negotiate on the part of the United States. Low noted that Kissinger "stated that as far as the White House was concerned, I had a completely free hand to negotiate in any area that was within NASA's overall responsibility. He stated that the president was in full support of these meetings and asked me to express the president's desire for cooperative efforts with the Soviet Union in space research and technology."[128]

In January 1971 Low traveled to Moscow to meet with Keldysh and the Soviet team "to discuss broader possibilities for cooperation." He was accompanied by Bill Anders, representing the White House, and several other NASA officials, including Arnold Frutkin, whom Low would appoint the overall project manager and person responsible for ASTP. Once in Moscow, Low admits that his latitude in negotiating directly on behalf of the president was a key element that allowed him to actually negotiate a specific agreement. Before returning to the United States, the so-called Keldysh-Low space cooperation agreement was signed. Low's authority made the signing possible, because he could say, "I'm coming here to appear at the request of President Nixon, who has conveyed his interest in these meetings and lends me his strong support."[129]

"I met Keldysh for the first time in Leningrad in May of 1970 at a COSPAR meeting," Low noted of his relationship with the Soviet Academy president.

We sat down and chatted awhile. And a docking mission didn't really come up during those discussions, but he was well aware of our interest in cooperating, and he told me his side was working very hard and perhaps by fall they would be ready with an agenda for us to get together. Throughout this entire period, there were two thrusts. One was the science and applications agreement, the Keldysh-Low agreement of January 1971, which was independent of the docking business. And on Keldysh-Low, we worked and hammered out an agreement where both sides could see what the other one was doing. During those discussions, I asked Keldysh for a private meeting. Frutkin and I had a frank meeting with him in his private office, around a table. I told him that we would be interested in doing a demonstration flight test with existing hardware. The Soviet side had, to my knowledge, not yet started thinking about doing anything other than something way-off in the future. We proposed to go ahead with an early demonstration using existing hardware. He said he could not commit, but it sounded intriguing to him.

Low continued,

By the next time our working group people got together—Frutkin, Gilruth, Glenn Lunney, etc.—they started hammering out the possibility of an experimental mission. At the time, the *Salyut* space station had flown, and it became an Apollo-*Salyut* mission. And from then on until April 1972, the work was being done on Apollo docking with the *Salyut*. It wasn't until I got to Moscow in April 1972 that they sprung on us the fact that they'd rather make it Apollo-Soyuz. We asked them why, and they reasoned that the modifications to the *Salyut* would be expensive and difficult, and probably more than they wanted to bite off.[130]

Low had suspicions that the reasons really involved reliability issues concerning the launch vehicle used on the *Salyut* versus the vehicle used for the Soyuz. "It certainly had not demonstrated its reliability as often, and I think it was a matter of their side having a better assurance of success if they were to use Soyuz."[131]

"From a design point of view, I had no qualms with the proposal, but we had not yet worked out the operational details. How do we talk to each other? How do we handle mission control? We didn't really know how they thought and worked. A lot would need to be hammered out. I said we needed to work these details out before we could commit the president to a summit agreement. It was hard to get them to write things down," he recalled. "We had to become quite firm in the negotiations."[132] Low, Frutkin, and Lunney spent several days talking with their Russian counterparts to formulate the broad outlines of an agreement. When shown the Soyuz capsule during a tour, Lunney recalled, Low amazed the Russian engineers by hopping inside. He instantly began comparing the control layout with that of Apollo, and he wanted to fly the simulation. "That's something their bosses never would have done," Lunney said. They wouldn't have gotten their hands dirty like that. And according to Lunney, it elevated Low in their eyes; this was a man who knew what he was doing.[133]

After a few days, the three men formalized the final details over a hamburger lunch at the U.S. Embassy. They borrowed a typewriter from the staff and went to an upstairs room, where Frutkin typed while Low and Lunney dictated the agreement. "We went back to our counterparts after lunch, and then it was clear sailing," Low recalled. There was some wordsmithing up until the end, of course. In fact, Low remembers getting a call from a State Department attaché in Moscow a few minutes before he was due on stage to give a speech in California, for his assistance in finalizing the wording. This was a few hours before the agreement was signed by President Nixon and Soviet premier Alexei Kosygin on 24 May 1972.[134]

Senator Proxmire and others argued against budget allocation to support ASTP. They thought it too much of a risk for the astronauts' safety. Some called it a publicity stunt, and others were concerned with potential negative consequences of technology transfer to the Soviets.

As a result, Low kept a keen eye on costs. "Each time we reviewed the costs for the U.S./USSR early docking experiment, these costs have climbed to new high levels," he wrote in his notes.[135] "The last Office of Manned Space Flight (OMSF) proposal including a mission and back-up with a full set of experiments . . . approached $400 million. After a number of discussions and after I had told the OMSF people that I would recommend against us undertaking this mission, we reached a new number of $225 million plus a $25 million contingency for a total of $250 million." While still expensive, the $250 million budget was easier for Low and Fletcher to sell to Congress than a figure over $400 million.

The risk issue was a bit of a sticky problem, as the Soviets had experienced loss of life on the *Soyuz 11* mission to the *Salyut* space station in 1971. The issue not only concerned Congress; it also concerned Low, the working groups, and the astronauts. It was a major focus of the team throughout the technical and operational negotiations of ASTP. Then, three months before ASTP launched, the Soviets experienced an aborted launch on *Soyuz 18*. The cosmonauts were fine, but it gave everyone pause. The aborted flight, Senator Proxmire said at the time, only "reinforces my deep concern that ASTP may be dangerous to American astronauts."[136] Of course, the irony of the U.S. concern around safety is that after ASTP splashdown, it would be mistakes made by the American crew within the U.S. command module that would expose them to harmful, poisonous gases once back on Earth. The Soviet team experienced no such health or risk issues, nor did either crew during the actual ASTP linkup on orbit.

As for the tech-transfer sensitivities, Low saw it much differently than the politicians. A focus on simplicity was one of his principal design philosophies in engineering—something that overeager engineers tend to forget, in their zeal to push the envelope on new technology. In an interview with *Time* magazine, he insisted that it was the U.S. that learned technological lessons from the Russians and not vice versa. Low told the magazine that the project exposed the NASA designers "of the sophisticated Apollo system" to the "functional simplicity of the less costly Soviet space hardware." Low's takeaway was simple, if not prophetic in light of the coming, complex shuttle hardware: "We have to learn not to overdo things when they don't have to be overdone."[137]

Low would ultimately travel to the Soviet Union five times in his personal efforts to get the ASTP mission off the ground. At times, his trips would be kept secret, like the one before the Nixon-Krushchev summit in 1972. Low had to buy and set up his travel arrangements himself, outside NASA channels, and only Fletcher, Mary R., and Low's secretary knew where he was during that trip. He enjoyed "the semi-clandestine nature of the trip" and said it "lent some excitement" to the normal, everyday office work he was doing in Washington.[138] Other times, Low was conducting high-level international diplomacy between the nations, such as when he presented a formal plaque to a group of cosmonauts at the Gagarin Museum in Moscow, commemorating the first man in space. "It was an emotional moment," Low wrote to Bob Gilruth, who designed the plaque. "It was obvious that they were pleased at the recognition by us of their being first in space."[139]

In October 1973, just two years away from liftoff, Low found himself back in Moscow to continue negotiations to advance systems safety, communications, and the minute operational details necessary to undertake such a joint exercise. In the moment, the Soviet Union finally started to lift the veil of secrecy on its own operations. In the process, the hopes first expressed by Low at the Langley tour in 1959 finally came to fruition. "George Low and his aides became the first foreigners to visit the main Soviet space mission control center outside of Moscow," the *New York Times* reported. "Heretofore, the only foreigners to visit Baikonur were Presidents de Gaulle and Pompidou of France. The warm treatment of Mr. Low and a team of American specialists . . . was read as a deliberate gesture by Moscow to emphasize its interest in Soviet-American cooperation."[140]

At a news conference in Baikonur, he expressed confidence in the mission and its ability to be achieved on schedule. "An impressive amount of work has been accomplished," Low said. "The progress made, and the quality of the joint work to date gives us high confidence that the scheduled launch date can be met." In addition, Low reported that both the Soviet and the American crews "were making headway in learning each other's language." He also commented on the growing closeness of the American and Soviet teams, noting "a very dramatic change in our relationships" since the project was initiated.[141]

Still, most of the trips remained packed with back-to-back meetings designed to hammer out all the technical, logistical, and political requirements necessary for the two Cold War enemy nations to link up in space. It would take five years of dedicated, concentrated effort on the part of Low, and the rest of the team, to finally accomplish the mission.

The day before Apollo CSM 111 took off from Cape Kennedy on 15 July 1975, Low was interviewed by the *New York Times*. "Apollo-Soyuz is a very difficult and a very important step in space exploration," he declared. "It is a pioneering venture of a different kind—pioneering in international cooperation. We are learning how to work with our chief competitor; how to cooperate on a most complex project where the final report card will be visible to all. We are opening the door for more cooperative efforts in the future."[142]

"The public reaction to the flight appeared to be extremely good," Low wrote in his personal notes. "In reading through all of the issues of 'Current News' for the duration of the flight, I would say there were about 90% positive stories, and 10% negative stories. Considering everything that NASA does, that is a very high percentage of pluses."[143] ASTP was a public relations coup. The mission was broadcast around the world. For the first time, citizens of the Soviet Union watched a launch from Baikonur as it happened. In the United States, Low singled out the efforts of John Donnelly, for his public affairs work on the project, as well as Bob Shafer, for his role in coordinating the international television broadcasts—including the actual linkup and the handshake itself, between the American astronauts and the Soviet cosmonauts.

It was also Apollo's coda.

"The party's over," mission control radioed up to the ASTP crew on their last day on orbit. "Time to come home."[144]

George Low, the man who helped put America on the path toward the moon from his earliest days in the space business under Abe Silverstein in Cleveland, saw the program through to completion. In a way, it was his spaceflight coda too.

"I was asked in many interviews whether I wasn't sad that Apollo was finally over, and that the Apollo system would fly no more," Low wrote in his notes in summation of the Apollo-Soyuz Test Project mission. "My answer generally was that I might be nostalgic, but not sad. Apollo and its hardware had been extremely good, and that carrying on beyond Apollo with the same hardware into *Skylab* and ASTP was also good. But this didn't mean that we should fly the same 1960 hardware indefinitely. I was nostalgic that I no longer had my model A Ford, but I really wasn't sad that we stayed with the model A technology forever. Space people are forward-looking people. Finishing ASTP now gives us an opportunity to devote full time to the Shuttle and things beyond. The end of ASTP was an occasion to be happy about the future, rather than sad about the past."[145]

6

Giving Back to the Future

The bicentennial year started off in a surreal moment of praise for NASA from one of its biggest critics in Congress. "An impressive record of cost cutting that could well set an example for many agencies has been achieved by NASA since the mid-1960s," Sen. William Proxmire said on 1 January 1976.[1] He indicated that NASA had made its tight budget go further by "making one dollar do the job that it took two dollars." Perhaps the best steward of the government's money, NASA was still perceived by many in the general public as just being a waste of money. Its prolonged austere budgeting did not come without consequences on both programs and personnel. Low was struggling to hire the appropriate qualified people to fill critical positions, and he was also losing his top talent at an accelerated pace. People just couldn't afford to work for NASA.

"With continuing inflation, and no relief in sight insofar as an executive pay raise is concerned, our people problems are getting more and more severe," he noted in his daily diary. One deputy associate administrator "decided that he could no longer afford to stay with the government," especially with the cost of living as it was in Washington, and he had just announced his resignation. Low's longtime friend and colleague Ed Cortright told Low that he "is reasonably sure that he will leave NASA after Viking is launched for a job in private industry at three times his present salary."[2]

Low tried to bring the issue to the attention of the Ford White House, holding meetings with Dick Cheney, who provided economic counsel to the president, and Donald Rumsfeld, who was currently Ford's chief of staff. It was a new administration, and Low was hoping to be able to make some progress on improving NASA's compensation policies. "I visited the White House to discuss the executive pay situation, and the meeting can best be summed up by saying we got a great deal of sympathy, but no real help."[3] Indeed, Ford's first budget request gave NASA a 4 percent raise, to $3.6 billion. But that was well below the rate of inflation, and NASA had to once again postpone the purchase of a shuttle orbiter by one year. At the same time, a worrying sign was starting to develop that the United States was losing ground on its international competitiveness. "U.S. world-wide predominance in science and technology had eroded over the past 15 years," a major National Science Foundation report announced. "Based on a review of 492 major technological innovations, the U.S. share of the total sank from 75 percent in the 1953 to 1965 period, to just 58 percent in the 1971 to 1973 period."[4]

Many within NASA's management ranks were finding better opportunities outside government. Low was not immune to the situation. In fact, he had been approached by the White House concerning his interest in heading up the Federal Aviation Administration (FAA) in the months preceding ASTP. At the beginning of the Ford administration, all federal agency heads were asked to tender their resignations, pending reapproval by the White House. Fletcher and Low both tendered theirs; they were quickly reappointed. In the process, the head of the FAA, Alexander Butterfield, who had revealed the existence of the White House taping system during the Watergate investigation, was not. He left his post on 31 March 1975.

In April 1975 Low was approached by Bill Coleman, President Ford's secretary of transportation, and interviewed for the job. The initial interview with Coleman was positive. "Good first meeting," Low wrote. "He says the right things about the relationship, but I would need a handshake agreement on no political interference, as well as my authority to do the job, including reorganizing, and hiring of personnel." He had concerns that the job would be tough to manage without a strong deputy, as George Abbey had been for him during Apollo or as he served for Fletcher, Gilruth, and Silverstein. "Will I be able to handle both the inside management job and the external affairs job?" he asked himself. "No—unless I have a good deputy, who can carry out the day-to-day operations of the organization." The FAA faced huge challenges, with "poor morale, bad organization, and tied up in bureaucratic lethargy." There was also a lot of undue political influence, no different from what NASA was experiencing. It would make the job more difficult if he did not have a free hand. "It needs to be streamlined, reorganized, and reinvigorated," he noted. Key to the process would be to "cement the relationship between diverse outside groups—congress, pilots, controllers, airports, carriers, manufacturers, all for the purpose of running a smooth, safe, economical aviation operation for the United States."[5]

The next day, Low called Frank Borman, who was now chairman of Eastern Airlines. "Best news I've heard," Borman told him. Borman told Low that Low would have his unequivocal support. He also said the FAA was an archaic organization that was politically hamstrung and operationally ineffective. Most importantly, it drastically needed to upgrade and modernize its systems and technology. On relationships with key constituents, Borman said that the hardest part would be working with the pilots' union. This relationship would be Low's toughest, and he would need to gain their confidence. Borman offered to help in this regard, and he believed that Low could do it. "You are the right man for the job," he told Low.[6]

Low also called Bill Anders, who was now chairman of the Nuclear Regulatory Commission. Anders told him that the challenge would be greater at the FAA than at NASA. The issue at NASA was one of consolidation, whereas at the FAA it was one of sparking innovation—it would need a complete overhaul. If Low wanted to eventually pivot to private industry, Anders advised that the FAA job would give him a better base from which to make the leap. "It's a doable job," he said, "but the rate of improvement will be slow."[7] Evolution, not revolution.

After considerable thought, Low informed Fletcher on 24 April that he was going to formally throw his hat into the ring. "I told Fletcher that I was going to say yes to the FAA opportunity," he wrote. "My basic reason, I explained, was that I cannot turn my back on a challenge."[8] He had spent the last twenty-six years of his career at NASA. It was time for a change. Fletcher, for his part, told Low that he thought Low's chances of becoming NASA's administrator were good if he stuck around, but he also understood.[9] If Low were to get the FAA job, Fletcher would want him to stay on at NASA at least through ASTP. While Low disagreed with the need to stay on, he said he would consider it.

The competition for the FAA post came down to Low and former astronaut Pete Conrad, who was now working in private industry. Coleman pushed for Low. However, key players in the White House wanted Conrad. William N. Walker, a lawyer and director of the president's personnel office, also objected to Low. "Secretary Coleman believes that George Low is better qualified than Conrad to carry out this function," Walker wrote to President Nixon via Don Rumsfeld in a White House memo dated 28 April 1975. Walker wrote that

Coleman is "being urged to object to Conrad" but that Coleman would accept Rumsfeld's judgment. Walker then pointed out that Conrad was a Republican, while Low was an independent, and that Low had enemies on the Hill who called him "politically naïve"—without explaining what they meant by that, although Low's desire to have a "free hand" from politics in managing the FAA might explain this response. Walker also believed that Conrad's "demonstrated toughness" assured that he would "not get chewed up by the FAA bureaucracy."[10] They offered the job to Conrad, who, after a period of time, turned the job down out of fear of losing his military pension if he accepted.

"Sometime toward the end of June, Don Rumsfeld, who was then the President's closest assistant, called me while I was in New York and he told me that the President wanted me to become the Administrator of FAA," Low explained. "I told him that I would need several days to consider this, and he tried to apply pressure on the basis that the President needed an immediate answer." The nation's largest organization for transportation safety was now without a leader, and the Ford administration was starting to get criticism for leaving it open so long. Low pushed back, saying he needed time to discuss with Fletcher, his family, and others. "Several days later, Bill Coleman called me and . . . showed some renewed interest in my joining the FAA."[11] After a week of considering the pros and cons and discussing it with the people he trusted, Low turned the job down. The position would just be more of the same.

If he were to make a change, it would need to be outside government. As it was for the others at NASA, compensation was also an issue, and the silly political calculations back and forth on the FAA position solidified his thinking. He wrote in a letter,

> I believe the time has now come for me to begin to think about other opportunities outside of Government. In large measure this is due to the fact that Government pay scales have been frozen for six years, during a time when the cost of living has increased 45 percent. With three youngsters in college, and two more to go, and with no real hope that anything will be done about the Government's executive salaries, I simply cannot continue to consider staying with NASA for the indefinite future. However, this does not mean that I am in a great hurry to leave. I would only want to change jobs if I could find a new and even more difficult challenge.[12]

Through the latter half of 1975, he contemplated his next move. He made a list of executive recruiting firms, and he started making inquiries on the best approach for a job search. He made a list of one hundred people whom he would tap from his network of contacts to solicit ideas and recommendations. He also began the process of self-evaluation, listing out his strengths: "I have been successful at running things," he wrote. "Have a solid, practical engineering background; Good at organizing, managing, selecting people, belt tightening; I'm willing to make tough decisions, and I take responsibility for them; Innovative in planning and marketing, yet also practical; Know the aircraft industry and Washington; Have worked in the USSR." He had a lot to offer. He'd made up his mind. He would move on from NASA in 1976.

In the fall of 1975, a few months after ASTP, he noticed a mole on his back that needed attention.[13] It was diagnosed as late-stage malignant melanoma. He immediately had it surgically removed; the prognosis, however, was not good. On a referral from his colleague and friend Dr. Charles Berry, NASA's chief medical officer, he went for follow-up treatment at MD Anderson Cancer Center in Houston. He underwent an experimental course of BCG immunotherapy treatment that sought to stimulate his immune system to fight and prevent the spread of the cancer. The treatment appeared to be successful, and his melanoma retreated into remission.

He returned to work. But his focus had changed. By the end of the year, Low would begin the process of searching for a new position in earnest.

"George and Mary R. were both the same way—they were very much interested in the broader world," astronaut Rusty Schweickart explained. Having a wider view was the reason he rotated astronauts like Schweickart through NASA Headquarters assignments. It was often an eye-opening experience.

> I went up there to learn more what management was about. I wasn't interested in just flying the spacecraft as an astronaut; I also wanted to know how things get managed and organized. How do you deal with budgets? What is congressional testimony and all of that? I went to Headquarters to broaden my own experience and background. You realize it is a very different world. While being an astronaut is exciting, your missions were few and far between. You find that while working in Wash-

ington is nowhere near as exciting physically, you're in the process of deciding what gets done and what doesn't get done, what the priorities are, and creating new programs. That becomes its own reward, which I was frankly surprised to find became attractive.[14]

As NASA director of user affairs—a fancy name for space applications—Schweickart had been a liaison between Low and Jacques Cousteau, looking at ways that NASA and Cousteau could work together to promote earth sciences. "George knew that I had spent huge amounts of time underwater in the water immersion facility at MSC on the Skylab program. I was very much into scuba work. George himself loved scuba diving, and he had a natural affinity for Cousteau. He put us together," he said.[15]

Low was interested at the time in the work of people like author Stuart Brandt, editor of the *Whole Earth Catalog*, and physicist and space activist Gerard O'Neill, according to Schweickart. "George was not a narrow person, by any means. We would get into all kinds of conversations—about space colonies, the environment, and all kinds of things. George had this wonderful, broad interest in almost everything that was going on, especially anything that was on the leading edge, and people who were doing far-out things."[16] NASA had LANDSAT (formerly the Earth Resources Technology Satellite) used to photograph and image the world in order to monitor changes to the environment. Low and Cousteau were looking for ways to use LANDSAT with oceanic research.

"I spent a very enlightening Christmas week with Commander Cousteau on *Calypso*. It was Christmas a year ago, off a small island in Mexico," Low said, during a 1976 conversation with Schweickart, Cousteau, Cousteau's son Phillipe, and Stewart Brand. "We started talking about what we can jointly do in learning how to better understand the oceans and to begin to get on top of some of the problems. . . . In fact, we did a joint project last summer in the Caribbean to try to see whether satellites can help determine the shallow ocean depths, the depths of the ocean where ships tend to run aground and spew oil all over." The experiment was successful. Low found that the most intriguing aspect was the use of over thirty satellites in the process, from LANDSAT to navigation, weather, and communication satellites, all communicating with the instruments on the *Calypso* in real time. "It was a tremendous example of how modern tools of scientists can be put together to get a better understanding of this globe we live on. I think that's really the key to our total future. We've got to use all of these tools, whether they're on the ocean, whether they're under the ocean, whether they're in space, to help understand the fundamental factors that govern our environment, our life, and everything else."[17]

Low's interest in the broader world, and in the application of space science to that world, was a main reason Schweickart was not surprised by Low's desire to leave. The inspiration and innovation would need to come from the next generation, which would learn from and figure out new uses for the technology that the Apollo generation had created. "It was very easy to see with George's nature that opportunities outside of NASA would become a lot more interesting at this time than staying at Headquarters."[18]

During his job-search outreach, Low was most serious about two intriguing positions that opened up in early 1976. The first was to serve as president of his alma mater, Rensselaer Polytechnic Institute back in Troy, New York. Low had been serving as a trustee to the institute since 1971, and he was intimately familiar with its challenges and opportunities. It was in desperate need of renewal. The institute was at a crossroads, and without radical changes, it would face the real possibility of extinction. He was also approached to be the president of a start-up uranium-enrichment company as part of the Garrett Corporation, an industrial conglomerate based in Los Angeles, California. The high-tech start-up, called TRENCOR, would build and operate a centrifuge enrichment plant in Texas on behalf of the U.S. Energy Research and Development Administration. While the Garrett opportunity offered him enticing financial rewards, including rich bonus targets and generous stock grants, the unique opportunity at RPI represented something even more valuable to him—a challenge with a purpose.

"It was an opportunity to work with the faculty and the students to build for the future," he wrote.[19] "I am totally dedicated to the proposition that the future of our Nation depends on the advancement of science and technology, and I know of no better way to help assure that future than by contributing directly to the technological education of our young people."

NASA announced on 19 March 1976 that deputy administrator George Low would be leaving in June to become the fourteenth president of Rensselaer Polytechnic Institute. "I know of no other person who has contributed as much to achieve and maintain United States leadership in space exploration," Jim Fletcher said in the announcement. "George Low's influence on the manned space flight program from the very beginning of NASA, his courageous decisions in the Apollo program, his direction of the agency's programs toward the practical application of space technology, and his initiatives in the development of international cooperation in space have earned the admiration and gratitude of the entire country."[20]

"How do you say goodbye to an organization after you have given the best part of your life to it? I think the simple answer is that you don't," Low wrote in his notes, under a section titled "Farewell to NASA." "People give you gifts, people give you parties, people say nice words, and you respond in kind. But all of these are only surface symbols. Down deep, you will always wonder whether you did all that you could have and should have done for NASA and its programs. And down deep, you will always worry that your successors will not carry on in the tradition which you tried to establish."[21]

Upon Low's announcement, there was an outpouring of dismay from key people around NASA. Astronaut Joe Allen's reaction typified the response. Allen, a scientist-astronaut selected in 1967 who would go on to fly on the space shuttle, had replaced Gerry Griffin as head of NASA Legislative Affairs when Griffin rotated back down to Houston. Two weeks before the announcement of leaving government service, Low called Allen. He wanted Allen's help in executing a plan for the announcement.

Allen recalls,

George Low was one of the great leaders of the space program, without question, an extraordinary human being. I was flabbergasted and shocked. I could not believe he was going to leave. We had to notify Congress and many related VIPs. We had to draw up press releases, and I can remember I had a hard time doing that, because I was having to read these through tears in my eyes. I didn't want him to leave NASA. I was fearful about that, because he was such a brilliant man and such a leader. I had just a bad feeling in the pit of my stomach. We've been through some catastrophes, and we're going to have more in the future. It's just too non-forgiving an environment without extremely smart people involved.[22]

In a post–George Low NASA, people worried that "the intrusion of politics and bureaucracy would compromise the performance of the agency" and that with the culture and character change of new people—outside, non-technical people—coming on board in executive management, the attention to detail and the ability to air out engineering issues and balance differing technical opinions would be lost.[23] It was a similar concern that Low himself had expressed to young John Conway many years ago in his cubicle—it was the leadership that set the tone, that often made the life or death difference at NASA.

"In looking back over 26 and a half years of service, I can honestly say that NASA is the best Agency in government," he told the NASA center directors on 21 April 1976, in his stylistically pragmatic and blunt manner. "Even though NASA continues to have an outstanding reputation for results, people view it as being in search of a mission, and that far too many people are in self-preservation mode. You need to overcome this, or plan on continuing to lose 5 percent of your budget every year. There is no harm in letting your best people go." At the same time, he cautioned them not to lose sight of the need to continue to innovate, believing that great things still were ahead for NASA in the 1980s with the shuttle. Low encouraged them to continue to focus on bringing in young talent and to expand NASA's diversity of ideas through the EEO program. "It is not easy to leave NASA, but the Agency's prospects for the future remain excellent," he said.[24]

At his formal NASA farewell dinner on 26 May 1976, he struck an equally pragmatic and positive tone—his message, one of hope, was also a not-so-subtle challenge:

> You must know that I am leaving NASA with mixed emotions. I am sad for all of the things that I will not be a part of—the tremendous opportunities that lie ahead in aeronautics and space. At the same time, I am looking forward to my new career at Rensselaer, particularly the challenge to contribute directly to the technological education of our young people. I will not make a long speech this evening—but I do want to leave you with a message. The message is simply this: technology and innovation have made America a great nation in the past, and only technology and innovation can keep us great in the future.
>
> But technology needs a pacesetter, and that pacesetter has been, and should continue to be, our activities in aeronautics and space. Pacesetting is not easy. There will be those who will stand in the way of prog-

ress, who would rather say no than yes, who will have you analyze and study a new idea to its death, who will have you spend more time justifying than doing productive work; those who just plain do not have the vision of the future that you and I have.

Don't let them discourage you. As long as you believe in what you are doing, as long as you know that what you are doing is right for the future of humanity, keep pushing ahead no matter how high the hurdles.

Keep setting the pace.[25]

***

The announcement of Low's appointment as president of Rensselaer was generally met with excitement both on campus and in the city of Troy, which had developed a strained relationship with the institution over the preceding years. His appointment was the result of an extensive search conducted by a committee of faculty, students, trustees, and staff, who reviewed more than 340 nominations for the position.[26] Low and two others were forwarded on to the board of trustees for consideration. "He was my number one candidate," said a student member of the selection committee, highlighting Low's capacity and desire to really "listen to both students and faculty" on the future of the institution. "It has been clear for some time that RPI has had problems communicating with its faculty and with the community in which it thrives," the *Troy Times Record* editorial board said in assessing the importance of Low's appointment, especially in light of his predecessor's reluctance to get involved in the community at large. "These have been serious problems that many of us who admire RPI looked on with great concern. . . . We hope that George Low will bridge the gap."[27] A few days later, editor Joe Cooley wrote about his admiration for Low's approach to the appointment. Saying he wanted to repay RPI by serving because it is where he got his start in life, Cooley quoted him as saying, "It may sound corny, but it is true." To which Cooley replied, "Low's sentiment doesn't sound corny, because it's just what RPI and Troy need."[28]

Before giving his formal inaugural as president in October 1976 and taking up his office the following June, he visited the campus and Troy several times for meetings, planning sessions, and attempts to open up the much-needed lines of communication. In his first appearance on campus, he held not only a press conference but multiple meetings with faculty, students, and staff. He also engaged in an active listening campaign, refusing to hint at what needed to be done on campus and insisting that he'd "rather spend a bit more time listening to other people and getting the best opinions from all of the people, from the students, the faculty, the members of the board, and then, when I am on campus, decide how we can all best work together."[29] At his press conference, he struck a similar note: "I am not president yet," he said. "I am not going to state my goals or issue them until I've talked to everyone in greater detail. I've a lot to learn."[30]

"He appears to be taking our pulse," a student at the press conference commented, seemingly impressed with his interest in getting to know the students.[31]

"I think we are in very good shape right now in terms of physical plant," Low said, when discussing the state of the campus. "We're building a new Engineering Center, we have a new library, we have the Communications Center, we have all sorts of buildings here that will make this, when it is completed, one of the best equipped engineering campuses that I know of. What's left to be done, then, is to provide the best possible education within that. It will be up to all of us to work together to use this new facility for the best possible education."[32]

After walking the campus and talking to more groups of students, Low went to the Sage Dining Hall for lunch. It was a simple gesture, but it clearly broke with past tradition and was met with warm acceptance. "He is the first President to eat in the faculty dining room in a decade," said one delightedly surprised professor.[33] "I met with student leaders earlier today," Low told the editorial board of the school newspaper. "They suggested I ought to eat in the student dining hall every once in a while. I think that is a great idea, and I intend to do it."[34] Not only did Low do it, but he also went out of his way to go to clambakes, dances, picnics, dinners, lunches, and all kinds of school activities to engage and interact with students. During the opening of hockey season, when students camped out in long lines for days to get access to the best seats in the rink, Low would walk up and down the student lines serving them juice and doughnuts for breakfast. It's the same management approach he took all his life. "The important thing is that you work for people, and you try and get the best out of people, and that means talking to people, understanding their problems and helping solve them. That's what I've done all my life."[35]

In one of his last NASA personal notes, he wrote of his visits to campus, describing them as a "fairly busy time." He was pleased that "the outlook on campus about my Presidency is still overwhelmingly good," but he cautioned himself not to get too optimistic. "I have not yet gotten involved in the day-to-day, but I will be faced with problems as soon as I arrive." Several key faculty and administration people had announced their resignations, independently of Low's becoming president. They were all going on to "better schools" and "better opportunities" that RPI could not match. In addition, Low noted, "There is a serious problem concerning computers on campus," and he also had a long list of priorities that would need to be tackled based on his time attending board meetings. He was already formulating a plan in his mind that would culminate in his ambitious "Rensselaer 2000" long-range plan. Like NASA, RPI was desperately in need of shoring up its mission and putting forward a strong, motivating vision for the future. Private institutions like RPI were suffering in the stagflation economy of the 1970s, with reduced budgets and staffing, while constantly having to increase tuition rates. "The first few months will be months of work," he said, work to understand fully the needs of the various constituents on campus, in the community, and in the region at large. "It's a new challenge for me. I've been away from here for a long time. I've not been a college president before. I've done new things, so I am not afraid of it, but the only way I know how to do the best job possible for RPI is to meet with people, listen to people, and then let them participate in some way in where we decide to go from here."[36]

Low was installed as RPI's fourteenth president on 8 October 1976, at the age of fifty. In his inaugural address, he struck an optimistic tone about the future and pushed back against the risk-averse, technophobic naysayers in society at large. Under his leadership, Low wanted Rensselaer—like his last challenge to the people of NASA—to set the pace, to be the trendsetter of the future, not just in engineering, but in society at large:

> How could being at Rensselaer even begin to compare with going to the moon? The answer to that question is simply this: Here is where the future lies. . . . Imagination has brought us the wisdom and the adventures of our past, has let us set our foot on the moon and our minds on the stars. . . . Imagination gives us the hope for the future, a sense of direction, but it takes much more to reach our destination. It takes knowledge, and courage, and confidence—above all, the confidence to

be willing to take risks. Yet today much of our society has an aversion to risk; people are afraid of science and technology, of where our experiments might lead us; afraid of the unknown and of the future. But if we are unwilling to venture, there can be no hope: without risk, there can be no progress. We now need a positive force, a move to transcend our physical image, our preoccupation with the negative. We need a place where the creative potentiality of the individual can develop and grow—to cope with the unexpected, to face the coming crises of our time. I believe that Rensselaer should be that place, should provide that force. At Rensselaer, we can apply our imagination to achieve the impossible, we can apply our knowledge to look to tomorrow with confidence.[37]

He further outlined his vision for the future of the school, pointing out its traditions that need to be continued—molding and developing problem solvers and providing an education that prepares its students "to meet the challenges and problems of a rapidly changing world"—as well as those areas that need to expand, especially in graduate studies and research in the "fields which we believe will be of prime significance in the world of tomorrow." His vision for the students included not only a foundational education but also a transformative one. They should learn not to just build bridges out of iron and steel but also "bridges that span from now to the future; bridges which span between science, engineering, and technology, on the one hand, and society on the other." It is only through a more rounded, socially relevant education that technically oriented people become valuable, because only then can they "transcend that background and develop insight into the nature of people and society."[38]

His address was stirring and inspirational, and it touched at the heart of an institution that was in desperate need of renewal. "As for the immediate problems facing Rensselaer, President Low feels that an improvement in the school's long-range planning tops the list," wrote Mark Tortorici, in a summary of his address in the *Polytechnic*, RPI's campus newspaper. "In other words, what should be done to insure the college will be better than it is, or will it still be around fifteen or twenty years from now—what more should be provided, how much more can Rensselaer grow."[39]

Equally as enthusiastic was the *Polytechnic*'s editorial board. They met his openness and engagement with similar encouragement:

Already we have seen a good deal of the man, his ideas and actions, and we are favorably impressed. His most striking difference from our last president is his openness. Low makes himself accessible to all factions of the campus community. During the summer, he circulated a memo with his home phone numbers on it to all faculty and staff members, urging them to call in the event of a problem. Indeed, he has even returned phone calls made by staffers of this publication, something that would have been unheard of a year ago. . . . [He] seems to be pleasing just about everybody. This is no mean feat, considering that the faculty, administration and students often do not see eye to eye. . . . Sensing another immediate need, Low has put together a Presidential Task Force to study possible solutions to the apparently inadequate computing situation here. The last president merely told us that the computer was good enough . . .

President Low, we like your enthusiasm and we like your work.[40]

Early the following year, he held a meeting with RPI faculty and staff in which he discussed the current state of the school, as well as his short-term and long-term plans. As for the current state, he used a stoplight report that was used in Apollo. "In the space program, when we reported status of programs or projects, we started out with a color-coded alert—green (good shape), yellow (minor problems), and red (critical problems). Using this convention, I would report RPI in 1977 as definitely green." He cited the outstanding faculty, the growing research in significant areas of scholarship, the addition of new facilities, and the financially sound balance sheet. "In short," he said, "Rensselaer is on the move." This attitude, he cautioned, was unusual among institutions of higher learning, most of whom where mired in retrenchment rather than building—an attitude of doom and gloom rather than vitality. The reason for Rensselaer's positive outlook? "We have a dynamic board excited about RPI's future, who are knowledgeable about the need for engineers and scientists in our nation's future, and who are willing to bet on that future." In the end, however, just being positive wouldn't be enough. They needed to be right about their ideas, and they needed to be realistic in their aspirations. He called on the faculty and staff to work with him on his Rensselaer 2000 long-range plan, in the spirit of being able to "take on things which fit our style and plan and reset those which others might do better than us." The goals were to establish broad but realistic objectives for the university and to "develop the means to achieve those objectives in the years between 1982 and 2000." This dynamic process would stretch out for more than a year, including input not only from them but also from the students, community, and other interested outside constituents, like alumni and the businesses that employed RPI graduates.[41]

While the long-term plan would be collaborative, Low identified, in the short-term, a handful of immediate needs. "There are some things which we clearly must do in the very near term—things which are now on the agenda—to which we must be committed." Included among them were the needs to provide an up-to-date computing system that would put RPI in the lead among schools in the nation; to hire in key open positions, such as a vice president for administration and budget and a vice president of student affairs; to revamp the engineering curriculum; and to successfully complete the new campus facilities, as well as a very real need to support affirmative action. "We have been giving lip service in this area, with no real action," he noted. "It is an area where I am going to start to get tough." He underscored the need for more minorities and women on the faculty and staff, as well as a need for more minority and women students. "Not one of you is doing your job as faculty, staff, or administrators, if you are not doing your best in affirmative action," he warned. "I will consider your performance in this regard as much as I will consider your performance in other areas."[42]

Low's philosophy for the future of RPI was inspired and informed by the work of Frederick Terman at Stanford.[43] A trained chemist and engineer who did his undergraduate work at Stanford and his graduate work in engineering at MIT, Terman is widely credited with being the father of Silicon Valley. As the dean of engineering at Stanford after World War II, he spearheaded the creation of the Stanford Industrial Park, in which the university leased portions of its vacant land to high-tech and start-up firms. Companies such as Hewlett-Packard, Eastman Kodak, General Electric, and Lockheed all moved into the park, "making the mid-Peninsula area into a hotbed of innovation which eventually became known as Silicon Valley."[44]

"When we set out to create a community of technical scholars in Silicon Valley, there wasn't much here and the rest of the world looked awfully big. Now a lot of the rest of the world is here," Terman said of the success of his industrial research park.[45] Low wanted to replicate this model at RPI on vacant land that the university owned but could never develop. He felt it would create a magnet for innovation and job creation for the region and would also provide employment, internships, and research opportunities for RPI students and faculty. From his Apollo days, he knew the valuable connection between government-sponsored programs (like the space program) with private industry (such as contractors like Rockwell and Grumman) and research universities (such as MIT). He wanted RPI to do for the region what Stanford did for Silicon Valley or what Harvard and MIT did for Route 128 in Boston, which *Business Week* dubbed "America's Technology Highway." His goal was to build a technology park for RPI, one with high-tech regional companies working in both industrial and biotechnical sciences. Additionally, he wanted to attract start-up firms to leverage the natural tension between corporate-funded research and university research. Universities generally develop research that is published for all to use; corporations generally want to patent their technology and keep close control on it. How to entice the output-oriented and private corporate interest, while at the same time engaging the knowledge-oriented expertise of RPI's researchers? Low's solution was to build the nation's first university-business incubator, in which university researchers were encouraged to develop and market commercial applications for their work. It would give birth, he firmly believed, not only to attracting world-class faculty, students, and researchers to RPI but also to new industry, jobs, and prosperity for the capital region, which included Albany, Schenectady, and Troy. It was his moon shot for RPI. And it succeeded beyond anyone's wildest expectations.

Under his leadership, RPI began the development on a three-hundred-acre technology park on university land in nearby North Greenbush, New York.[46] With the money Low raised from the State of New York and from private corporations such as GE, IBM, Kodak, and Colt Industries, RPI began construction on the nine-story, two-hundred-thousand-square-foot, $60 million Center for Industrial Innovation on the Troy campus. It would become a hub of international research in the fields of computer graphics and information technology, manufacturing technology, and integrated electronics. By the spring of 1984, National Semiconductor became the first company to sign a long-term lease. By 2009 the park would have over fifteen buildings and house over seventy high-tech firms employing more than 2,400 people. RPI, the State of New York, and IBM would partner in 2007 on a $100 million nanotechnology innovation center at the park.

At the ground breaking for the technology park, Low spoke about the future results from what would be built on the site:

> With this groundbreaking will come a major step to make the capital district a great center for business and industry in the United States. We are marking a new beginning of a new era for high technology companies in the city of Troy . . . at RPI, already eight new companies have been born within the past year. Soon this will be a place where universities, industry, and government come together for the greater benefit of all. And all will benefit. RPI will benefit because this site will be an extension of our laboratories; a place where students and faculty can be exposed to entrepreneurial ventures; American Industry will benefit because in this environment, they will be able to flourish and grow; The People of the Capital District and the State of New York will benefit, because the project will bring commerce and trade and employment. . . . It is an American dream, funded with private capital, for the benefit of the public good. Soon this will be a place where people come to work—first hundreds, then thousands of them—to develop high technology products, to revitalize American productivity, and to continue to rebuild the economy of the State of New York.[47]

Over the course of his presidency, Low would become known as an activist leader in the best sense of the word. The style and leadership that he developed at NASA came out in full force as he worked to propel RPI into national prominence. He insisted on dedication, hard work, high energy, and excellence in execution—no different from what he expected from his people at NASA. "His name inspires memories: of his gentle manner, his workaholic habits, his love of running, and of his notes in green ink—'green hornets,' they were called—which darted from his pen when he thought a subordinate could do a better job," wrote William F. Hammond Jr. in the Rensselaer alumni magazine. "People credit him with revitalizing the college, making it a center of research, graduate education, and technology entrepreneurship."[48]

Among his many accomplishments as president, Low elevated the university to national prominence in computer technology, with the installation of a state-of-the-art facility in the center of campus that was housed in an old, unused, but stately former church—a symbolic gesture toward the importance of computing and technology for the future. He developed and implemented his Rensselaer 2000 plan, which established realistic and achievable goals for the university over the next two decades, including expanding the school's graduate programs, hiring world-class faculty, focusing on the commercialization and technical application of research, and overhauling the School of Engineering to propel it into nationally prominent leadership. He more than tripled the money spent on research, from $6 million in the 1970s to over $21 million by 1984. And he started and completed a major fund-raising campaign that exceeded its original goal by 37 percent, pulling in more than $50 million to help finance the school's expansion via private means.[49]

More importantly, he brought along the energy, drive, and inspiration for everyone at RPI to pull together and make the most of their hard work. He set the tone. He became the pacesetter. "George wants this to be one of the best institutions in the world," said Lee Wilcox, vice provost for student affairs, to a reporter in 1982, during the middle of Low's tenure. The reporter noted that people on campus had never worked harder yet still seemed to enjoy every minute of it. Wilcox attributes this infectious spirit of excellence to Low: "It rubs off on people. His desire for quality is so omnipresent that you don't dare do a sloppy job."[50]

In February 1979 Low traveled to Washington DC for a national education conference, scheduled to take place during the week of 19 February.[51] As fate would have it, the entire Eastern Seaboard would get hit with a surprise snowstorm that dumped over two feet of snow on DC. The whole town shut down, and for the first time in its history, the Smithsonian Air and Space Museum would be closed to the public for several days. Holed up at his hotel, Low decided to venture out into the storm to get some fresh air and visit the museum. He made his way across the mall to the Smithsonian. Even though it was closed, he rapped on the door. Since he had been there so many times and was so well known for his role at NASA, they let him in. He wandered through the silent, empty display rooms and made his way to the *Apollo 11* command module known as *Columbia*. He spent a long time there, gazing into the machine he helped build—thinking back on all that had transpired, all that had been achieved. His eyes drifted from one toggle switch to the next. He could tell the story of each one, where it was made, what it did, when it was needed, what the backup plan was if it didn't work—everything. It must have warmed him to stand there and go back in time in his mind, back to when the future of space exploration was just starting to open up for NASA and the country. When he was deputy administrator, his family lived in Virginia. Their house had a living room with a rounded row of windows at the top. At night when he could not sleep—when something was on his mind or bothering him—he would get up and sit in a chair and look up at the stars through the windows. He'd pick a star and watch it move through the night sky, marking time. Who knows—maybe he did the same thing as a child out on the deck of the *Veendam*. In either event, here he was now, doing something similar but marking time through his mind, working the switches and traveling the miles of wires inside *Columbia*, reliving the adventure from a perspective that no one else could have or ever would have. It was a rare moment to revel in the past for a man who spent most of his time focused on the future. He surely must have given thought to the day when the shuttle would finally take flight, when more people would be able to venture off planet, when engineers and scientists would begin to push the boundaries and start moving farther out into deep space. Apollo might have been in the past, but its effort was more than just the evolution of technology. It was also a revolutionary leap into the future.

During his eight years in office as president of RPI, Low spent a great deal of time speaking on campus, in the region, and throughout the country to focus a spotlight on national issues that were important to him, including American industrial competitiveness, national productivity improvements, and advancements in technology policy and applications, among many other themes of the 1980s. A key theme of his lectures had to do with what he viewed as the dangers of a risk-averse society and an increasingly polarized government of special interest groups that prey on that risk aversion to halt forward progress in society, technology, the environment, and human well-being.

In the Roy V. Wright Lecture, presented to the American Society of Mechanical Engineers in Chicago on 19 November 1980, he drove these points home in a long, powerful keynote titled "Productivity, Professionalism, and the Public Interest." He started out by reminding the audience that in 1930, Wright wrote an important essay in which he "admonished the engineer to assume a responsible role as a member of society." Low cautioned that "economic stability and growth alone will not guarantee that we will once again become a leader among nations" and that the true strength of the United States, both at home and abroad, "depends first on the power of our ideas" and then on "the ability to do what we value and cherish on this earth." The United States, from Low's perspective, was living in a time of small-mindedness, without courageous goals and plans—just "national policies based on hope and wishfulness." But as engineers, he said, "you know that rhetoric, hope, and wishfulness are not enough, that results depend on doing, that accomplishments depend on action and performance."[52]

It was a cathartic speech, one seemingly built on his personal experience of decades of engagement (and frustration) in Washington as he watched the triumphs of NASA diminish within the halls of government and in the minds of the public.

Could America ever hope to achieve such big things as a nation again? It was a good question, one he wanted the engineers to ponder: "There has been a dramatic change in the way we govern ourselves. Twenty years ago, when we started a program called Apollo, I knew a Washington that could get things done. Ten years ago, after we landed on the moon, I saw a perceptible slowing in how we did things. And now, for many reasons, the gears of government are grinding to a halt. Today, I wonder whether we could start another Apollo, much less accomplish it."[53]

He cautioned the group that hyperpolarization and minute focus on special interests over the cause of the common good threaten the fabric of U.S. politics and society. It was a classic forest-for-the-trees speech, acknowledging that "in our search for progress we soon forgot that costs grow exponentially as we try to approach absolutes." Without risk, there can be no progress. Without forward movement—even if it's imperfect or might fail—we become static and inflexible, and our overall quality of life and standing in the world starts to decline. As engineers, it was not only their job but also their responsibility to get engaged in the broader, bigger picture of society.

Acknowledging that there is no better form of government than the one in the United States, he warned—based on his experience of trying to save NASA in a post-Apollo world—that "the elected leadership can no longer establish a strong sense of direction, nor can it impose the discipline to follow the course promised by its political party. . . . The result is disorder and confusion, wherein the special interest wins at the expense of the general interest." He did not, however, want to be misunderstood:

I strongly support the need to make continued progress toward our societal goals. I will fight for affirmative action, for protection of health and safety on the job, for clean air and water, for human rights and human liberties. But I also know that to achieve these goals, we need the underpinning of a stable economy. Without regaining that, as a first priority, our progress in all other areas will not only come to a halt—we will lose ground. The only way that I know to reestablish economic growth and to make steady progress toward all of our goals is to do so in a planned and measured way, optimized in the general interest of our nation. Continued contradictory policy and continued suboptimization for each of our special interests can only cause us to fail in all of our objectives.[54]

While Low was serving as RPI president, the nation also turned to him for important national committee work. During the Jimmy Carter administration, for example, he was called on to review the nation's general aviation safety procedures. During President-elect Ronald Reagan's transition period, Low was called on to serve as chairman of the NASA transition team, making broad recommendations for the future president's space policy. A committed public servant, Low always accepted the call, irrespective of the office holder's political party affiliation. If he could help—especially in areas that concerned aeronautics, space, or education—Low was all in.

On 21 December 1979 the National Research Council, on behalf of the transportation secretary, tapped him to chair a thirteen-person committee to investigate and evaluate the Federal Aviation Administration's regulations and certification process for aircraft design, production, and maintenance, in the shadow of the tragic crash of a DC-10 jet during takeoff in Chicago on 25 May 1979.[55] Due to pylon-mount damage, the left engine separated from the plane's wing during takeoff, causing the plan to crash, killing 271 people. The six-month, independent investigation was not designed to review or evaluate the earlier crash; rather, Low's committee was "to review the overall FAA airworthiness certification procedures of the agency" and to question "the efficacy of the nation's system for assuring both the traveling public and domestic and foreign purchasers that American built aircraft continue to warrant their worldwide reputation for safety, durability, and reliability."[56]

Low and his committee held public hearings in Washington and visited the aircraft production facilities of Boeing, McDonnell Douglas, and Lockheed; the two regional field offices of the FAA designated to review and certify aircraft for airworthiness; and the maintenance departments and facilities of three airlines, including Air Florida, United Airlines, and USAir. They also took briefings from aviation officials from Great Britain and France, as well as from various U.S. institutions, including the congressional subcommittees of the National Transportation Safety Board, NASA, and Transportation.[57] Low's report did not find anything "that should undermine public confidence in today's commercial air travel." It did, however, make recommendations and lead to major changes at the FAA to improve confidence in the overall safety of the U.S. aviation industry.

"The FAA engineering staff is today considerably less competent than the engineers in the industry they regulate," Low said at a press conference in Washington upon delivering his report, titled "Improving Aircraft Safety."[58] The panel recommended that the FAA overhaul and strengthen standards intended to allow safe landings of damaged planes—a rule change that might have prevented the Chicago DC-10 crash, according to Low. He recommended a more competent technical staff for the FAA, as well as a centralization of its engineering functions, which were "scattered across the country." Noting that there had been five FAA chiefs in the last ten years, he underscored the need for more continuity and stability within FAA management. Transportation Secretary Neil Goldschmidt, who accepted the report, said he would order "an immediate analysis of the panel's findings with the intent of putting into effect promptly any recommendations that will improve our procedures."[59]

As a result of the report's findings and the secretary's review of the recommendations, the FAA initiated development of the Aviation Safety Analysis System in order to leverage new computer technology to support safety oversight and decision-making. The system reduced the burden of paperwork, freeing up inspectors to spend more time in the field actually investigating safety protocols.[60]

On 12 November 1980, shortly after Ronald Reagan was elected president, Voyager 1 flew within seventy-seven thousand miles of the planet Saturn. The small, robotic spacecraft began sending dramatic images back to Earth from over 950 million miles away. Scientists were stunned. The high-resolution images "showed a world that seemed to confound all known laws of physics."

Its six rings turned out to number in the hundreds, and they "appeared to dance, buckle, and interlock in ways never thought possible."[61]

Low, in the early 1970s, helped save Voyager out of the budgetary constraints that caused NASA to cancel the Grand Tour robotic program to survey the outer planets. Since then, NASA had suffered under continued budget cuts and indifferent White House space policies. The nation was also in the grips of a recession, and U.S. industrial productivity was suffering at the expense of increased productivity and higher quality from European and Asian competition, especially from Japan. Inspired by the stark contrast between the amazing response to the Voyager images and the lack of support for the U.S. space program, and its seeming connection with America's industrial slump, Scripps-Howard science writer Don Kirkman urged President-elect Reagan to "take a look at NASA's faltering program," in an article titled "America Has Begun to Turn Its Back on the Stars." Noting that no American astronaut "has ventured into space for five years," Kirkman argued that the budget cuts and canceled programs have meant that the "technological discoveries that contributed so much to the nation's industry during the 1960s are now few and far between." Kirkman spoke to John Glenn, now a senator from Ohio, who eviscerated the prior administrations for "robbing the country" of a space capability "that was unmatched in the world." Jim Fletcher, now no longer with NASA, commented that NASA "has long hankered to begin imaginative and innovate new space ventures" but that "the last four presidents have firmly squelched every proposal."[62]

Low, when reached by Kirkman, did not shy away, either, in his criticisms, which were published on 15 November 1980. It was a theme he continued to hammer home in his speeches and in his writing since before leaving NASA. "The old space program was one of the greatest things that ever happened to our country," he observed. "It had a little bit of everything: manned flights, planetary exploration, science and applications. If we don't excite the people about the moon and the stars, they won't get excited about making the country productive again. I'm afraid the marvelous mission to Saturn was the last hurrah of America's space program unless we decide to do something to revive it." He pointed out that the Soviet Union's current space program was far bigger than the one in the United States, having launched five satellites for every one that the United States put into orbit, as well as "a significant number of manned flights to their permanently orbiting space station." He underscored, however, that "the Russians do not have the technological expertise to carry out a mission like the one the United States flew past Saturn." There was still time to stop the slide and recapture America's preeminence in space. Turning to the hope of more funding and a renewed interest in space with the president-elect, Low laid out a not-so-subtle challenge: "Reagan has the same opportunity President Kennedy had in the 1960s. He can lead this country in a new direction, and part of the answer could be a bigger space program."[63]

A few days later, Reagan's transition office announced the appointment of Low to head up his advisory and transition team on NASA.[64] Heading to Washington once more—this time only temporarily, despite some rumors and articles suggesting that he might be a replacement for outgoing administrator Robert Frosch—Low would spend the next four weeks working five days a week with NASA and the transition team to come up with a comprehensive report and suggestions for the president. Before Low's team issued their report, the outgoing administrator, Frosch, lent his stamp of support: "The transition team is a very strong, professional, space-knowledgeable team," and he hoped their efforts would spark the Reagan administration to set a strong, national goal in space. Frosch lamented the lack of support for NASA and said he struggled to recognize what a burden it was for the agency not to have strong goals. "I did not succeed in finding some kind of single, 'Let's go to the moon' kind of goal. It took me a long time to realize what people meant when they talked about setting goals for the agency," Frosch said.[65]

On 19 December 1980 Low delivered his *Report of the Transition Team, National Aeronautics and Space Administration* to Richard Fairbanks, the director of Reagan's Transition Resources and Development Group. In his note to Fairbanks, he pointed out that the report "presents a balanced view of the status of the agency, its problems, strengths, and potentials," and that NASA represented for Reagan "an opportunity for positive accomplishment." In contrast to other government agencies that were mired in controversy and problems, "NASA can be many things in the future—the best in American accomplishment and inspiration."[66] True to his word, the transition report pulled no punches:

> In 1958, the people of the United States set out to lead the world in space. By 1970, they had achieved their goal. Men walked on the moon, scientific satellites opened new windows to the universe, and communications

satellites and new technologies brought economic return. With these came new knowledge and ideas, a sense of pride, and national prestige. In 1980, by contrast, United States leadership and preeminence are seriously threatened and measurably eroded. . . . In recent years, the agency has been underfunded, without purpose or direction. The new administration finds NASA at a crossroads, with possible moves toward either retrenchment or growth.[67]

The team made ten specific recommendations for dealing with NASA's issues, including requesting that the administration make a strong statement of support for the agency, as well as develop a specific purpose and direction for the space program in conjunction with the flight of the shuttle in the spring of 1981. While the president was a strong supporter of the military applications for space initiatives, "space during the Reagan presidency was not a top priority issue—at least, not before the *Challenger* accident." Still, he "gave the civilian space program just enough priority and budgetary support to allow it to move forward at a modest pace." Among the new starts and initiatives that he did approve was the go-ahead for the space station as "a highly visible symbol of cooperation among the United States and its 'friends and allies.'"[68] While not the grand, visionary goal that NASA craved, the Reagan administration's eventual space policy represented a victory of sorts for Low. In his original 1961 Low committee report on the feasibility of a lunar landing within a decade, he had stated, "An important element in the manned spaceflight program is the establishment of a space station in earth orbit. Present thinking indicates that such a station can be established in the same time-period as manned lunar landings can be made, and also that many of the same technological developments are required for both. Although both missions were broadly considered in planning developments for the lunar program, only the lunar requirements are discussed in this paper."[69]

Low's work on the Reagan transition team (and his behind-the-scenes advocacy) helped to finally resurrect his original recommendation. When satellite deployment was considered off-limits after the *Challenger* accident, NASA would gain a new mission—building the International Space Station.

He was also highly active in various professional and industrial organizations, including being an influential board member of General Electric. In fact, he was a member of the board that selected the legendary Jack Welch to become CEO of GE back in 1981. "George Low was a very intelligent, detail-oriented person," Welch said. "He was never really the first person to speak up in any meeting. He often listened first, but you knew he was paying attention. And when he spoke, it was always something important that cut to the chase, something valuable that you wanted to hear."[70]

After being installed as CEO, Welch invited Low to present at Welch's first corporate officers meeting. On 6 October 1981 Low flew down to Phoenix, Arizona, to present on the subject of the human side of quality—a topic he strongly believed in. During the worst days after the *Apollo 1* fire, he mentioned to journalist Robert Sherrod that one of the problems of Apollo management before he took over was that the other people didn't get as emotionally invested in the effort as he or Gilruth or Kraft or Slayton would. It was the human spark that provided the magic over the purely clinical technical judgment. Before speaking, he thanked Welch for bringing him under the hood to the event, which he found both stimulating and informative. "To meet with all of the officers of GE, and to put names and faces together, I am leaving this evening with a very good feeling." Turing to Welch, he said, "Jack, I think the company is in good hands."[71]

"The human side of quality this morning is 140 highly motivated General Electric officers looking at their Japanese watches to see how long their speaker is going to keep them away from this afternoon's pursuits of higher productivity," he said, to laughs. Drawing on the recent first spaceflight of the shuttle *Columbia* (which President Reagan hailed as a return to the country feeling "as giants once again" and feeling "the surge of pride that comes from knowing that we are the first and we are the best") and his experiences in Apollo (in which quality and productivity had suffered to the point that it cost three precious lives), Low lectured on the importance of quality and productivity and how they were keys to the pursuit of making either a nation like America or a corporation like GE "first and best." He pointed out that productivity is simply a measure of efficiency—of output over input, the number of widgets per person per hour. "It depends on many things: on how well we plan and manage, on the tools we invest in, on the use of new techniques and technologies, on innovation, and on the quality and skills of the people who do the work." As for quality, he went back to both the space program and the current national preoccupation with Japanese ascendency in global trade. "What do you think of, when you think of quality? Certainly, the space program, the flawless flight and pinpoint landing of *Columbia*. Certainly, also, the Japanese watches I referred to before, the Japanese cameras used by your General Electric photographers, and the Japanese cars so many of you own. In fact, with the exception of the space program, and perhaps also with the exception of aircraft, computers, and household appliances, 'Made in Japan' has replaced 'Made in the United States' as the world-renowned symbol of quality."

He wasn't just speaking of consumer goods. He also referred to the decline in quality of most things in everyday American life:

> music and art; of the letters we write and the books we read; of air and water; of everything we do and feel and touch. . . . When I think of quality, I think of reaching for the best, of investing effort and skill to produce the finest possible result, of attaining the highest standard, always within the available resources. . . . You can do it well, or you can do it half-well. And when you do it half-well, there is no quality; when you have to do it twice, there is no productivity. . . . Oh yes, quality has indeed declined.

Quality can be renewed, he told the GE officers, for it is a very human thing. It is predicated on people and their attitudes; change and inspire them, and you can change and inspire a renewal in quality and productivity. "Quality," he said, "is the key to American renewal, to the reestablishment of the United States as a leader in world competition, and hence it is the conduit to our economic well-being."

To underscore his perspective, he used examples from Apollo. Driving home the human aspect of quality, he talked about the implied promise of good work.

> In our quest for the moon, we could not tolerate less than the best. How do you motivate people to give their very best, always? In Apollo, we sent the astronauts to visit the people who did the work, to go out on the factory floor even during the night shift, with one simple message: "Please do good work, my life depends on it." Suddenly that which had been abstract became real. . . . In 1972, we were testing the last of the giant Saturn rocket stages, the one for *Apollo 17*. Every man and woman working on that stage knew that when the job was done, they

would get their pink slips—they would be out of work. Yet that *Apollo 17* stage was delivered to Florida without a flaw, the best of all the stages we tested. Why? Because those who worked on it wanted to show that they could do the unexpected, that they were proud of their jobs, that they weren't about to be sloppy just because everybody thought they would do less than the best.

He named a number of Apollo examples to discuss the importance of employee incentives, good planning, and execution, as well as the importance of looking out into the future to stay ahead of the curve—especially amid a rapidly changing technology landscape.

In closing, he brought the moon and the stars back down to the present. In recalling a graduation ceremony at the RPI field house the year prior, he talked of standing on the stage in his cap and gown, conferring degrees on 1,753 new graduates:

> As I watched their eager faces, as I heard name after name called out, I thought about quality and motivation—my motivation. I remember another time and another place—the morning of an Apollo flight to the moon, breakfast with the astronauts, small talk, and finally, the hand-shake, and my saying to them: "I have done my very best, from now on it is up to you." I suddenly realized that commencement—the scene before my eyes—was just another morning of a launch, that once again I was saying: "I have done my very best, from now on, it is up to you." The moment of truth. What is your launch morning, your commencement, your moment of truth? Doesn't everybody have one, if only we will help them find it?
>
> That is the human side of quality.

***

In February 1984 Low's melanoma aggressively returned.[72] He went back to MD Anderson to receive treatment. A steady stream of his former Houston colleagues would come by to visit, including Griffin, Abbey, and John Young, among many others. "I'd take John Young with me to visit George," Abbey recalled. "The two of us would go down and visit with him. He was optimistic, as always. I can remember one visit when he was telling us that when he got out of the hospital this time, that he was going to take up golf—that's where a lot of people make decisions, out on the golf course. 'I think I'm going to take up golf,' he said. He must've seen something in John's reaction, because he said, 'Don't worry, John. I'm not really going to be taking up golf.'" That's the kind of guy he was, Abbey explained, always thinking of other people.[73]

Abbey and Young were there to privately inform him that his son George David was going to be selected as a shuttle astronaut. The official announcement was going to be made in a few weeks. They weren't sure he was going to make it long enough to hear it, and they wanted to reassure him that absolutely no favoritism was involved. "I wanted him to know that we selected David on the merits of his ability, not because he was George Low's son," Abbey said. "We could tell that he was pleased. It meant a great deal to him."

Unfortunately, the cancer had metastasized, and the treatment was ineffective this time. His prognosis was terminal. He returned to Troy—to his family and the work he loved. He could be seen on the job as president of Rensselaer—working hard, holding meetings, interacting with students, and making bold plans for the school's future. So focused and dedicated on planning for the future into his final days, he became uncharacteristically annoyed at the board of trustee's decision to rename the school's Center for Industrial Innovation after him. After all, that's what you do when people die; you name things after them—like the Kennedy Space Center or the Johnson Space Center. He was upset not at the kind gesture but by its implied meaning: he was going to die. "He kept coming into the office, even though he was getting worse and worse and worse with the cancer," Shirley Malloy, Low's longtime assistant at NASA and, for a short time, at Rensselaer, explained to a reporter.[74] Of course, back at the office is exactly where one would expect him to be—visiting the shop floor, getting his hands dirty, working hard, asking the tough questions, paying attention to the details. Fifteen years and a day after the launch of *Apollo 11*, George Michael Low passed away quietly at home with family at his side on 17 July 1984. He was just fifty-eight years old.

The next day, the White House announced that he would be awarded the Presidential Medal of Freedom for his contributions to education and the nation's space program.[75] Mary R. would accept the medal on his behalf the following year. On 20 July 1984 President Reagan signed a bill to declare the day Space

Exploration Day in honor of the fifteenth anniversary of the return of *Apollo 11*. In the presence of the assembled press; government dignitaries; NASA officials; and Neil Armstrong, Buzz Aldrin, and Michael Collins, he paused the ceremony to say a few words about Low. After recounting his many contributions at both NASA and RPI, the president paused. Pen in hand, before signing the bill, he said, "We're grateful for what George Low has done and the ideas he stood for. We shall miss him very much."[76]

Three days later, New York senator Patrick Moynihan introduced in the Senate Low's extensive obituary from the *New York Times* for the record. Addressing the Senate president and all his colleagues, a solemn Moynihan paid moving tribute to Low not only as a citizen of his state but as a great American and engineer. "George Low was one of the great engineers of this age," he proclaimed. "He died much too soon, much too early in a career that had demonstrated the extraordinary range of achievement of which he was capable. Mr. President, we have seen the death within this last fortnight of a man who might properly be described as the greatest engineer the world has ever known."[77]

Epilogue

In 2006 the tenacious Spirit Rover was placed in survival mode on the Martian surface. The intrepid robotic rover had lost use of its right front wheel just one hundred meters shy of its intended winter "safe spot." The wheel broke down in soft, sandy terrain that made maneuvering difficult. Getting Spirit unstuck, mission controllers backed away and decided to winter the hobbled explorer on firmer ground. The spot? Low Ridge, named in honor of George Low and situated just a few meters from the rover's current location.

Covered in finely layered bedrock, blocks of vesicular basalt, and a very salt-rich soil, Low Ridge would provide Spirit a target-rich environment for the long winter. "Low Ridge is actually a beautiful spot. It's one of the most scenic places we've been at," said Steve Squyres of Cornell University, the lead scientist on the Mars Exploration Rover project. "There's a lot of good science nearby."[1] According to Squyres, the team had decided to name major topographic features on Mars after some of the key pioneers of the space age. "There's a Von Braun Hill and a Goddard Hill," he explained. "Three big outcrops are named after Oberth, Korolev, and Tscilokovsky. Another outcrop is named after Max Faget. George Low was an obvious person to be included on this list. It wasn't difficult to conclude that he should be among the pioneers."[2]

Of all the places that could be found to help save and extend Spirit's mission over the harsh Martian winter, it was fitting that a place named after Low would be selected. Even if in name only, he was still helping to rescue NASA programs in the twenty-first century.

While the general public often thinks of the astronauts first when asked to identify America's pioneering space heroes, NASA engineers and scientists, such as those working on the Martian Rover program, invariably name a consistent handful of individuals—fellow engineers and scientists like George Low.[3] As legendary GE chairman Jack Welch once said, Low's legacy stands tall among his peers in his unique ability as a manager to knit together both people and technology, "whether in business or education, or in leading people in building a spaceship to the moon."[4] He was an extraordinary leader—full stop. "George Low was the epitome of an excellent manager," former astronaut and Eastern Airlines CEO Frank Borman said, echoing Welch's observation. "There is no question about it—I adopted many of his methods when I was CEO at Eastern."[5]

When looking back over the remarkable story that was George Low's life, a set of enduring leadership and management principles emerge. It is from these principles that Low's well-earned reputation and legacy originates. You can see these principles in action as they are consistent threads throughout his life. They are grounded in his tireless work ethic, obsessive attention to detail, and personal approach to management and leadership. They are not only worth noting but also, as Frank Borman did, worth emulating.

Low's top 10 management principles are as follows:

1. Pay excruciating attention to detail.
2. Focus on human-centered design and management principles.
3. Learn to listen and to ask the right questions.
4. Avoid the smartest-person-in-the-room fallacy.
5. Hire smart people so you can delegate, but support them with the right organizational structure, oversight, and work environment.
6. Never delegate your own decision-making responsibilities.
7. Manage with radical transparency, and be willing to admit your own mistakes. When making a mistake, learn from it; do not repeat it.
8. Take responsible, calculated risks.
9. Always lead by example, work hard, and never be afraid of going out onto the shop floor and getting your hands dirty.
10. Focus on the team—not just individual contributors—and lead with a compelling vision and sense of urgency, but don't ever forget to also have fun and enjoy what you are doing.

As a manager, leader, and educator, George Low made an indelible impact. It is hard to fathom where NASA or the state of U.S. technology and productivity would be if not for the amazing contributions of Low over the course of his fifty-eight short years on this earth. Indeed, one wonders just how much further along and productive our technological and industrial base would be if he had lived longer and could have continued his influence on both the national and international stage. If Low had stayed at NASA, one even wonders if the *Challenger* and *Columbia* accidents would have happened as they did, given the issues identified in postcrash analyses around engineering and management decision-making toward safety and acceptable risk.

Would he have been appointed NASA administrator? Would he have allowed the shuttle's operational costs "to eat NASA hollow," as Bill Anders described it; or given his focus on reducing costs and budgets and his willingness to pivot to newer technologies and systems, would he have taken preemptive action like he did with Apollo, when he realized the shuttle was not living up to Mathematica's overly optimistic projections? It's impossible—even futile—to speculate or even posit such hypotheticals.

But they do boggle the mind with the possibilities.

Even more futile would be to attempt to ledger up the impact he had on the individual lives of the countless tens of thousands of people who worked with and for him—whether at the NACA, NASA, RPI, or in private industry or the highest levels of the federal government. Of course, his contributions were not ignored during his lifetime. He won numerous managerial, science, and engineering awards and honors; chaired many blue-ribbon national panels; and was awarded a bevy of honorary doctorate degrees. Still, it was never about the rewards or the accolades with Low; rather, he was fueled by the challenges and needs of the time and by the fun he derived out of doing what he enjoyed most—being a dirty-hands engineer.

As a visionary, he worked tirelessly to see to it that humanity extended its reach beyond Earth's boundaries and was a pioneer and leading force in the early development of space travel. He was equally tireless in his desire for people to recognize the great contributions to society that NASA's work produced. "We do not throw money away," Low once said in the 1970s during a time when many questioned the value of national expenditures on the space program. "We spend it on Earth, and it feeds back into the economy. We develop high technology and increase the productivity of almost everything."[6]

Low was a passionate and unabashed futurist who believed that science and technology were key to the advancement of society, that technologists would be the bridge builders of the twenty-first century.[7] It was fitting, then, when NASA named its premier quality and performance award the George M. Low Award in 1990. Richard H. Truly, former shuttle astronaut and NASA administrator at the time, said in the agency's announcement, "George Low represented quality and excellence like few others."[8] Or when the New York State Center for Industrial Innovation—a center he helped to envision and create, along with the help of IBM, GE, and other leading New York State technology companies—was renamed the George M. Low Center for Industrial Innovation in 1984, the year he died. The center has been instrumental in turning the region into a miniature Silicon Valley, with advancements in computer science, the biosciences, additive manufacturing, and nanotechnology. "He was a true American technological leader," said RPI president Shirley Ann Jackson, at the dedication of the George M. Low Gallery at the George M. Low Center for Industrial Innovation on the Troy, New York, campus.[9]

Such is Low's legacy, not only for NASA and Rensselaer but also for America and the world.

Simply put, he was the ultimate engineer.

Acknowledgments

I am grateful for the help and resources of NASA's History Office—specifically from William P. Barry, Colin Fries, Stephen J. Garber, Jane Odom, Richard D. Spencer, and Elizabeth H. Suckow—as well as the help of Bob Jacobs, deputy associate administrator in the NASA Office of Communications. At the NASA John Glenn Research Center, I extend a big thank you to Anne K. Mills, records manager and history officer. I also want to thank Mal Peterson, Al Koller, and Norm Chaffee of the NASA Alumni League for their help in putting me in contact with folks who worked with Low during his NACA and NASA career. Special thanks to Ed Hengeveld for his expert NASA photo archive assistance.

This book would not be possible without access to the voluminous George M. Low Papers at the Rensselaer Institute Archives and Special Collections office of RPI. Special thanks to John Dojka, Jenifer Monger, and Tammy Gobert for their help and diligence on several research visits. Equal thanks to Chris Hunter, vice president of collections at the Museum of Innovation and Science in Schenectady, New York, which houses the General Electric archives.

A humble shout-out to my friends and fellow writers and space enthusiasts who provided encouragement, input, perspective (many via long phone calls and endless emails), and advice; read chapters; and made suggestions and who were, frankly, just there for me when I needed them: Alan Andres, Leslie Cantwell, Andy Chaikin, Francis French, Larry McGlynn, Bruce Moody, Chris Orwoll, Robert Pearlman, Jason Rubin, David Meerman Scott, Art Siemientkowski, Robert Stone, and Steve Worth.

I am highly appreciative and indebted to all those who sat for long interviews, told me their stories and experiences, exchanged emails with me, and gave so freely of their memories and recollections, including George Abbey, Bill Anders, Bob Blue, Frank Borman, Jerry Bostick, Andy Chaikin, John Conway, Gerry Griffin, Chris Kraft, Roger Launius, John Logsdon, Jim Lovell, Glynn Lunney, Jim McDivitt, Dorothy Reynolds, Walter Robb, Rusty Schweickart, Tom Stafford, Doug Ward, and Jack Welch.

Special thanks also to the Low family: Mark Low, Diane Murphy, John Low, Nancy Sullivan, and Eva Verplank. They spent considerable time going through their own archives, talking to me, meeting with me, and dealing with my seemingly endless email questions and follow up, as well as providing valuable insight into various drafts and edits of the manuscript. Thank you for your trust, honesty, and willingness to share the amazing story of your father and family with me and the world.

No list of acknowledgments for an Outward Odyssey series book would be complete without thanking Colin Burgess—a man whose passion and knowledge about space history are only exceeded by his genuine humanity, kindness, and talent. Also, many thanks to the staff at the University of Nebraska Press, especially Tish Fobben, Courtney Ochsner, Sara Springsteen, Rob Taylor, and Anna Weir. I also want to thank Jeremy Hall for his expert and detailed copyediting work; and Steve Flanagan for proofreading. You folks are wonderful, and I appreciate all your help, patience, and expertise.

Finally, I would like to thank my family—Karin, Tanja, and Philip—for their support and encouragement throughout the almost four-year process of writing this book, as well as their sacrifice of family vacations and weekends so that I could go on my research trips and spend time writing. You are the stars of my universe. Love is not a big enough word: *goi-goi*.

Appendix

Georg Löw was born in Vienna, Austria, on 10 June 1926. He came to the United States in 1940 and became a naturalized citizen in 1945, when he Americanized his name to George Michael Low. During World War II, he served in the U.S. Army from 1944 to 1946.

He attended Rensselaer Polytechnic Institute, Troy, New York, receiving a bachelor of aeronautical engineering degree in 1948 and a master of science degree in aeronautical engineering in 1950. In 1949 he married the former Mary Ruth McNamara of Troy, New York. And they had five children: Mark, Diane, David, John, and Nancy.

Professional Career

Low began his professional career by joining the National Advisory Committee for Aeronautics (NACA) as an aeronautical research scientist in 1949. At the NACA's Lewis Flight Propulsion Laboratory in Cleveland, Ohio (now NASA's Lewis Research Center), he specialized in experimental and theoretical research in the fields of heat transfer, boundary layer flows, and internal aerodynamics and published many reports in these fields. While at the Lewis facility, he was named head of the Fluid Mechanics Section and, later, chief of the Special Projects Branch.

In October 1958, when NASA was organized, he transferred to its Headquarters in Washington as chief of manned spaceflight. In 1960 he was chairman of the special committee that formulated the original plans for the Apollo manned lunar landing. He later became deputy associate administrator for manned spaceflight, where he was responsible for the management of the Gemini and Apollo programs and the field centers directly associated with those programs.

In February 1964 Low transferred to NASA's Manned Spacecraft Center (now Johnson Space Center) in Houston, Texas, as deputy director. As the center's general manager, he had overall responsibility for the Gemini and Apollo spacecraft efforts, as well as future program development, flight operations, and flight crew operations.

In April 1967, after the Apollo fire, he was named manager of the Apollo spacecraft program. Under his leadership, Apollo was redesigned and made flightworthy. To accelerate Apollo's timetable, he brought *Apollo 8* into the program as humanity's first flight to lunar orbit. Under his direction, five manned flights were flown, including *Apollo 11*, the first manned lunar landing in July 1969.

In December 1969 Low was appointed deputy administrator of NASA by President Nixon. As NASA's general manager, he had lead responsibility for internal activities, including planning; budgeting; financial, technical, and program management; and procurement. He guided NASA in the transition to the new goals of the 1970s and beyond, provided leadership in the space shuttle planning, and negotiated the space agreements with the Soviet Union, which laid the foundation for the Apollo-Soyuz flight and other joint space projects. He served as acting administrator of NASA from September 1970 to May 1971. After the appointment of Jim Fletcher as administrator, Low returned to his role as deputy administrator from May 1971 until his retirement in 1976, when he returned to the Rensselaer Polytechnic Institute, his alma mater, as its fourteenth president.

In July 1984 the White House announced that Low had been awarded the Presidential Medal of Freedom for his contributions to education and the nation's space program.

Professional Activities

He was a member and council member of the National Academy of Engineering; a fellow of the American Academy of Arts and Sciences and an honorary fellow in the American Institute of Aeronautics and Astronautics; a member of the Sigma Xi, Sigma Gamma Tau, and Tau Beta Pi honor societies; a trustee of the Hartford Graduate Center and of the Committee for Economic Development; and a member of the Board of Directors of the General Electric Company. Low was also a member of the Presidential Commission on Industrial Competitiveness and was chairman of the National Academy of Sciences, National Academy of Engineering, and Institute of Medicine's Committee on Science, Engineering, and Public Policy. Low was vice president of the Association of Colleges and Universities of the State of New York. He held a professional engineering license from the state of New York. Low was a member of the Explorers Club.

Honorary Degrees

Doctor of Engineering, Rensselaer Polytechnic Institute—1969.

Doctor of Science, University of Florida—1969.

Doctor of Engineering, Lehigh University—1979.

Doctor of Law, Hartwick College—1981.

Doctor of Humane Letters, Villanova University—1982.

Doctor of Science Honoris Causa, Albany Medical College— 1984.

Special Honors

NASA's Outstanding Leadership Medal, for his contributions to Project Mercury, 1962.

Arthur S. Flemming Award (Ten Outstanding Young Men in Government), 1963.

American Astronautical Society Space Flight Award, in acknowledgment of his contributions to the advancement of spaceflight and space science, 1963.

Paul T. Johns Trophy from the Arnold Air Society, for his outstanding contributions in aeronautics and astronautics, April 1969.

NASA Distinguished Service Medal, for his contributions to the *Apollo 8* manned lunar orbit mission, 1969.

NASA Distinguished Service Medal, for his contributions to the success of the Apollo program, October 1969.

Louis W. Hill Space Transportation Award from the American Institute of Aeronautics and Astronautics, for his leadership role in bringing the Apollo program to fulfillment, October 1969.

National Space Club Astronautics Engineer Award, for direction of the Apollo spacecraft program, March 1970.

National Space Club's Robert H. Goddard Memorial Trophy, for great achievement in advancing spaceflight programs contributing to U.S. leadership in astronautics and space, March 1973.

National Civil Service League's 1973 Career Service Award for Sustained Excellence, as the person most responsible for the success of the Apollo program, May 1973.

Rockefeller Public Service Award for administration, in recognition of distinguished service to the government of the United States and to the American people, December 1974.

National Academy of Engineering Founders Medal, for playing a central technical and management role in the U.S. space program, November 1978.

Austrian Cross of Honor for Science and Art, presented in recognition of his outstanding contributions in the area of space research and scientific progress, September 1980.

NASA Distinguished Service Medal, for his contributions to the space shuttle program, 1981.

American Association of Engineering Societies National Engineering Award, in recognition of his many contributions to mankind, May 1983.

University Foundation at Albany Citizen Laureate Award, given in recognition of his distinguished and sustained record of academic achievement, June 1983.

Capital District Business Review 1983 Executive of the Year, for his development of the Rensselaer Technology Park in North Greenbush, New York, and for his fundraising campaigns and improvements at Rensselaer Polytechnic Institute.

National Medal of Freedom and National Medal of Science, awarded posthumously by President Ronald Reagan, 1985.

Notes

"History looks different through everybody's eyes who has lived it," George Low said to a group of NASA historians, in an interview conducted by James M. Grimwood, Barton C. Hacker, and Peter J. Vorzimmer on 2 February 1967 (George M. Low Papers, 1930–1984). "I found out recently that Sorenson's history of Kennedy doesn't once mention space. Did you know that? Yet, to us, looking at Kennedy, from our point of view, if I were to give you a historic interview on Kennedy, I'd say that space was one of his major motivating forces. Yet his historian, personal and close to him, never mentioned it." In writing this book, I tried to focus on as many primary sources as possible—from Low's family and friends to the people who worked with him, as well as Low's own interviews and papers. I felt the burden to get his story right. To do that, I felt it necessary to quote him directly as much as possible, rather than to try to interpret his meaning. Occasionally, I needed to tweak a word here or there—never changing the meaning but perhaps the tense or a transition word to smooth out the narrative. This book is as much by him in that sense as it is about him. Low accomplished so much during his life and was involved in so many things at NASA that a book of this size cannot possibly capture every story and anecdote. Some might be disappointed in some of the stories I chose to leave on the cutting-room floor. Obviously, any sins of omission are mine alone. I do take comfort, however, in Low's words of caution to the historians—everyone has a different point of view, even those who lived through the same events but still end up seeing them differently. If you want to learn more about Low, please buy and read some of the wonderful books listed in the bibliography. Seek out the massive, insightful interviews of Robert Sherrod in the NASA History Office, and delve into the first-person accounts from his many interviews with Low. I also encourage you to go to Rensselaer Polytechnic Institute. Experience his legacy first hand. Tour the amazing George M. Low Gallery—filled with photos, documents, and historic artifacts—in the George M. Low Center for Industrial Innovation. And spend some time with his archives, especially reading his detailed "Notes for Dr. Gilruth" as Apollo program manager and his thoughtful "Personal Notes" as deputy administrator. It's like having a conversation with the man himself. They are a treasure trove of NASA history.

Introduction

1. George Low, interview by Robert Sherrod, 5 July 1972.
2. William Barry Furlong, "Biggest Headache in Space: Assignment—Fix Apollo," World Book Encyclopedia Science Service, 19 January 1968.
3. Walter Sullivan, "Preflight 'Crisis' Almost Routine: Attention to Detail on Land Helps Reduce Risks," *New York Times*, 27 May 1969.
4. George Low, quoted in the introductory 1984 Rensselaer Polytechnic Institute tribute video to George Low in the permanent exhibit on display at the George M. Low Gallery at the George M. Low Center for Industrial Innovation, Troy NY.
5. Furlong, "Biggest Headache in Space."
6. George Low, interview by Robert Sherrod, 21 June 1972.
7. Edward S. Goldstein, "A Conversation with Astronaut G. David Low," NASA, last updated 16 October 2008, http://www.nasa.gov/50th/50th_magazine /lessons.html.
8. Eva Verplanck, email correspondence with the author, 3 March 2016.

1. New Beginnings

1. Key details of George Low's early life in both Austria and the United States are amalgamated from a series of sources, including his official NASA job applications, security clearance forms, Low family records (from report cards to letters), local newspaper reports from the time, and interviews with Low's sister and children, as well as several interviews Low conducted with journalist Robert Sherrod in the 1970s.
2. The extensive details of the Löw family history in Austria are sourced from several correspondences with Low's children, including input from his sister Eva, as well as from the following key accounts published in their original German: Nowak, "Jüdische Unternehmer im österreichisch"; Tina Waltzer, "Schlossherren auf Zeit: Die Familie Löw in Matzen, Niederösterreich," DAVID: Jüdische Kulturzeitschrift, http://www.david.juden.at/kulturzeitschrift/61 -65/61-TinaWalzer.htm. Additional detailed information was also gleaned from several Holocaust Claims Resolution Tribunal case histories, including Österreichische Zuckerindustrie AG Syndicate, "In Re Assets of Gertrude Löw and Marianne Hamburger-Löw," 29 December 2006, case no. CV96-4849, Claims Resolution Tribunal database, http://www.crt-ii.org/_awards/ _apdfs/Low_Gertrude.pdf. For additional information on life in Vienna for nonobservant Jews and on the Anschluss, see Ciccone, "Personal and Political Advocacy"; Brook-Shepherd, *Anschluss*; Welzig, *Anschluss*.

3. Mark Low, email correspondence with the author, 28 January 2018.

4. Diane Murphy, email correspondence and phone interview with the author, 2016 and 2018.

5. ss *Veendam* Southampton departure manifest, 26 January 1940.

6. "They Changed Their Minds and Missed Death at Sea," *El Paso Herald-Post*, 12 February 1940.

7. "German Food Shortage Told by Relative of Local Man on Arrival in Ohio," *National Road Traveler News*, 22 February 1940.

8. "Somewhere over the Rainbow—Autobiography of Hans Georg Stern," *Gelsenzentrum*, June 2008, http://www.gelsenzentrum.de/story_hans_georg_stern .html. Stern, one of several hundred Jewish refugees housed in steerage, was aboard the *Veendam* on the same trip as Low and discusses what the journey was like—from the food they ate, to *The Wizard of Oz* film they watched, to life in London at the time of passage.

9. 1940s-era ss *Veendam* corporate marketing brochure, in the author's possession.

10. Shipping News, *Brooklyn Daily Eagle*, 5 February 1940.

11. Diane Murphy, interview by the author, 2016.

12. Weather, *Kingston Daily Freeman*, 5 February 1940.

13. Diane Murphy, interview by the author, 2016.

14. Eva Verplanck, email correspondence with the author, 13 March 2016.

15. ss *Veendam* customs declaration form and manifest, January 1940.

16. Diane Murphy, email correspondence with the author, 11 February 2018.

17. Eva Verplanck, email correspondence with the author, 31 March 2016, and Low family feedback on manuscript, email correspondence with the author, 28 January 2018.

18. Diane Murphy, interview by the author, 18 May 2016.

19. Goldstein, "Conversation with Astronaut G. David Low."

20. George Low, interview by Robert Sherrod, 5 July 1972.

21. Pete Dobinsky, "George Low Likes Initials—He Goes from NASA to RPI," *Press and Sun-Bulletin* (Binghampton NY), 30 May 1976.

22. Sylvia B. Kennick, archive catalog, George M. Low Papers, 1930–1984.

23. George Low, interview by Robert Sherrod, 5 July 1972.

24. George Low to Mr. James Slatter, 4 April 1966, George M. Low Papers, 1930–1984.

25. "Man behind Apollo—George Low," 12.

26. "Ricketts Building," Institute Archives and Special Collections, Rensselaer Libraries, Rensselaer Polytechnic Institute, http://www.lib.rpi.edu/Archives /buildings/ricketts.html.

27. George Low to Dr. Paul E. Hemke, 9 January 1969, George M. Low Papers, 1930–1984.

28. "Man behind Apollo," 12.

29. George Low, interview by Robert Sherrod, 5 July 1972.

30. George Low, RPI freshman orientation speech notes, 30 August 1980, George M. Low Papers, 1930–1984.

31. U.S. Civil Service Security Investigation Data Form, S6, For Sensitive Position, George Low, 10 December 1953, p. 2, George M. Low Papers, 1930–1984.

32. George Low, interview by Robert Sherrod, 5 July 1972.

33. "Zimmer Leads Draft Contingent," *Troy (NY) Record*, 11 November 1944.

34. "12 in Oneonta Area Called in Draft December 13: Men Will Assemble at the Palace Theatre at 8 am," *Binghamton Press and Sun-Bulletin*, 9 December 1944.

35. George Low, family letters, 1945, courtesy of Mark Low, 20 August 2016.

36. Mark Low, email correspondence with the author, 28 January 2018.

37. Mark Low, email correspondence with the author, 20 August 2016.

38. George Low, interview by Robert Sherrod, 5 July 1972.

39. Grathwol and Moorhus, *Building for Peace*.

40. Siegfried Beer, "The Soviet Occupation of Austria, 1945–1955," *Eurozine*, 24 May 2007.

41. Roosevelt, *On My Own*, 54.

42. George Low, interview by Robert Sherrod, 5 July 1972.

43. Crossfield and Blair, *Always Another Dawn*, 26.

44. George Low, interview by Robert Sherrod, 5 July 1972.

45. "Miss McNamara Wed to R.P.I. Graduate," *Troy (NY) Times Record*, 3 September 1949.

46. John S. Evvard, interview by Virginia P. Dawson, 14 June 1987, NASA History Office.

47. George M. Low, NACA pay stub, 12 November 1949, George M. Low Papers, 1930–1984.

48. For an extensive background on Silverstein and Lewis, see Dawson, *Engines and Innovation*.

49. Edgar M. Cortright, interview by Rich Dinkel, 20 August 1998, NASA Johnson Space Center Oral History Project.

50. George Low, interview by Robert Sherrod, 5 July 1972.

51. Roger Mike, "Learning from Past RPI Alum's Success," *Poly*, 14 September 2011, 6.

52. Bob Blue, interview by the author, 30 March 2016.

53. George Low, handwritten biographical notes, c. 1958, George M. Low Papers, 1930–1984.

54. Bob Blue, interview by the author, 30 March 2016, providing extensive detail about Low's work at the NACA at the time, his talents as a manager, his promotions, and his home life.

55. John Evvard, interview by Ginny Dawson, 14 June 1987, Glenn History Collection.

56. Bob Blue, interview by the author, 30 March 2016.

57. Bilstein, *Orders of Magnitude*, 33–36.

58. Bob Blue, interview by the author, 30 March 2016.

59. United Press International, "College Physics Instructor Arrested in Spy Case," *Great Falls Tribune*, 15 March 1951. For more background on events concerning the Rosenbergs and McCarthy's hunts for Communist spies, see also Radosh and Milton, *Rosenberg File*, 122–29.

60. Dawson, *Engines and Innovation*, 152.

61. James Hansen, *First Man*, 120.

62. James Hansen, *First Man*, 300.

63. James Hansen, *First Man*, 121–22.

64. Mark Low, interview by the author, 8 March 2016.

65. The description of Low's style is an amalgam taken from almost a dozen primary interviews with close work associates and colleagues who spanned his entire career, in which all of them—Bob Blue at the NACA, Gerry Griffin at NASA, Jack Welch at GE, and Dorothy Reynolds at RPI, just to name a few—would echo and amplify these same personality traits about Low and his style when describing him.

66. Dorothy Reynolds, interview by the author, 18 May 2016.

67. Bob Blue, interview by the author, 30 March 2016.

68. Nancy Sullivan, interview by the author, 31 March 2016.

69. Bob Blue, interview by the author, 30 March 2016; Furlong, "Biggest Headache in Space." This was also informed by various reflections from the Low children and by Edgar Cortright's speech at George Low's NASA farewell ceremony, 26 May 1976, George M. Low Papers, 1930–1984.

2. A Man in Space, Soonest

1. Furlong, "Biggest Headache in Space."

2. Dickson, *Sputnik*, 9.

3. Dickson, *Sputnik*, 14–15.

4. George Low, dedication and celebration speech for the GE Corporate Research and Development Center West Wing, Schenectady, New York, 4 October 1982, George M. Low Papers, 1930–1984.

5. Dickson, *Sputnik*, 17.

6. Brooks, Grimwood, and Swenson, *Chariots for Apollo*.

7. Dawson, *Engines and Innovation*, 149.

8. Virginia Dawson, "From Braunschweig to Ohio: Ernst Eckert and Government Heat Transfer Research," in Layton and Lienhard, *History of Heat Transfer*, 133.

9. McCurdy, *Inside NASA*, 73–74.

10. Dawson, *Engines and Innovation*, 150.

11. John Evvard, interview by Ginny Dawson, 14 June 1987, Glenn History Collection.

12. Dawson, *Engines and Innovation*, 150.

13. Warren J. North, interview by Summer Chick Bergen, 30 September 1998, NASA Johnson Space Center Oral History Project.

14. Swenson, Grimwood, and Alexander, *This New Ocean*, 63.

15. Von Ehrenfried, *Birth of NASA*, 36–40.

16. Dawson, *Engines and Innovation*, 158.

17. Swenson, Grimwood, and Alexander, *This New Ocean*, 75.

18. Link, *Space Medicine in Project Mercury*.

19. Swenson, Grimwood, and Alexander, *This New Ocean*, 75.

20. Dawson, *Engines and Innovation*, 163.

21. George Low, interview by Robert Sherrod, 5 July 1972.

22. Frank E. Rom, interview by Dr. Bernard Snyder, 10 July 1996, Air Force History Support Office, Bolling Air Force Base, Oral History Project, Glenn History Collection.

23. George Low, interview by Robert Sherrod, 5 July 1972.

24. George Low, interview by Robert Sherrod, 5 July 1972.

25. George Low, interview by Robert Sherrod, 5 July 1972.

26. Swenson, Grimwood, and Alexander, *This New Ocean*, 106.

27. George Low, STG interviews, 9 January 1969, 14 January 1969, 4 February 1969, transcript memos, George M. Low Papers, 1930–1984.

28. Andrew Chaikin, "Bob Gilruth, the Quite Force behind Apollo," *Air and Space Magazine*, February 2016.

29. T. K. Mattingly, interview by Rebecca Wright, 6 November 2001, NASA Johnson Space Center Oral History Project.

30. Abe Silverstein, tribute speech at George Low's NASA farewell ceremony, 26 May 1976, recording in author's possession.

31. Robert Gilruth, interview by Drs. David DeVorkin and John Mauer, 27 February 1987, NASM/SI Oral History Project, https://airandspace.si.edu/research/projects/oral-histories/transcpt/gilruth5.htm.

32. George Low, STG interviews, 9 January 1969, 14 January 1969, 4 February 1969, transcript memos, George M. Low Papers, 1930–1984.

33. Swenson, Grimwood, and Alexander, *This New Ocean*, 110–11.

34. George Low, speech, 10 October 1960, transcript by Walter T. Bonney, NASA History Office, George M. Low bio file #4133.

35. Michaelson, "The Tri-Service Program."

36. George M. Knauf to George Low, 9 May 1964, George M. Low Papers, 1930–1984.

37. Gilruth, "I Believe We Should Go to the Moon," 30.

38. Swenson, Grimwood, and Alexander, *This New Ocean*, 110–11.

39. Denise Grady, "George Low: From Spaceships to Scholarship," *Discover*, February 1982.

40. Swenson, Grimwood, and Alexander, *This New Ocean*, 111.

41. *Hearings before the Committee on Aeronautical and Space Sciences, Unites States Senate*, 86th Cong., 1st sess. (9 April 1959) (testimony of George M. Low, Chief, Manned Space Flight Program, NASA).

42. Logsdon, *Decision to Go to the Moon*, 56.

43. George Low, interview by Robert Sherrod, 5 July 1972.

44. George Low, interview by Robert Sherrod, 5 July 1972.

45. Bob Gilruth, tribute speech at George Low's NASA farewell ceremony, 26 May 1976.

46. George Low, interview by Robert Sherrod, 5 July 1972.

47. Bob Gilruth, tribute speech at George Low's NASA farewell ceremony, 26 May 1976.

48. George Low, interview by Robert Sherrod, 5 July 1972.

49. George Low, interview by Robert Sherrod, 5 July 1972.

50. Thomas, *Late Night Thoughts on Listening to Mahler's Ninth Symphony*, 47.

51. "In Memoriam: George M. Low," *Roundup*, 20 July 1984, 3–4.

52. Thomas O'Toole, "George Low: A Fervent Push for the Most Daring Venture," *Washington Post*, 20 July 1969.

53. Jerry Bostick, interview by the author, 10 March 2016.

54. "Man behind Apollo," 15; Furlong, "Biggest Headache in Space."

55. Furlong, "Biggest Headache in Space."

56. Sue Cronk, "What Would They Do without Their Wives?," *Washington Post*, 17 February 1963.

57. Furlong, "Biggest Headache in Space."

58. John Low, email correspondence with the author, 20 March 2016.

59. Furlong, "Biggest Headache in Space."

60. Dawson, *Engines and Innovation*, 164.

61. John Disher, interview with Robert Sherrod, 24 September 1970.

62. Swenson, Grimwood, and Alexander, *This New Ocean*, 116.

63. Rodney G. Rose, interview by Kevin M. Rusnak, 8 November 1999, NASA Johnson Space Center Oral History Project.

64. Furlong, "Biggest Headache in Space."

65. O'Toole, "George Low."

66. Low, "Project Mercury."

67. George Low, STG interviews, 9 January 1969, 14 January 1969, 4 February 1969, transcript memos, George M. Low Papers, 1930–1984.

68. Swanson, *Before This Decade Is Out*, 337.

69. Swanson, *Before This Decade Is Out*, 339.

70. George Low, interview by Robert Sherrod, 17 January 1970.

71. Wolfe, *The Right Stuff*, 62–63.

72. Burrows, *This New Ocean*, 289.

73. Wolfe, *The Right Stuff*, 64–65.

74. Wolfe, *The Right Stuff*, 65–66.

75. *Hearings before the Committee on Science and Astronautics, U.S. House of Representatives*, 86th Cong., 1st sess. (4 May 1959) (testimony of George Low).

76. *Hearings before the Committee on Science and Astronautics, U.S. House of Representatives*, 86th Cong., 1st sess. (4 May 1959) (testimony of George Low).

77. *Hearing before the Committee on Science and Astronautics, U.S. House of Representatives*, 86th Cong., executive sess. (28 May 1959) (testimony of George Low).

78. George Low, interview by James Burke, *Project Apollo*, BBC TV, 25 May 1979.

79. *Hearings before the Committee on Science and Astronautics, U.S. House of Representatives*, 86th Cong., 1st sess. (4 May 1959) (testimony of George Low).

80. *Hearings before the Committee on Science and Astronautics, U.S. House of Representatives*, 86th Cong., 1st sess. (4 May 1959) (testimony of George Low).

81. *Hearings before the Committee on Science and Astronautics, U.S. House of Representatives*, 86th Cong., 1st sess. (4 May 1959) (testimony of George Low).

82. *Hearings before the Committee on Science and Astronautics, U.S. House of Representatives*, 86th Cong., 1st sess. (4 May 1959) (testimony of George Low).

83. *Hearings before the Committee on Science and Astronautics, U.S. House of Representatives*, 86th Cong., 1st sess. (4 May 1959) (testimony of George Low).

84. Swenson, Grimwood, and Alexander, *This New Ocean*, 161.

85. Swenson, Grimwood, and Alexander, *This New Ocean*, 162.

86. *Hearings before the Committee on Science and Astronautics, U.S. House of Representatives*, 86th Cong., 1st sess. (4 May 1959) (testimony of George Low).

87. Foster, *Integrating Women into the Astronaut Corps*, 46.

88. *Hearings before the Committee on Science and Astronautics, U.S. House of Representatives*, 86th Cong., 2nd sess. (8 February 1960) (review of the space program).

89. Foster, *Integrating Women into the Astronaut Corps*, 46.

90. Gerry Griffin, interview by the author, 9 March 2017.

91. Glenn Waggoner, "The New R P I President Put Men on the Moon," *Troy (N Y) Times Record*, 23 March 1976.

92. Deke Slayton to George Low, 3 December 1975, George M. Low Papers, 1930–1984.

93. Foster, *Integrating Women into the Astronaut Corps*, 59.

94. *Hearings before the Committee on Science and Astronautics, U.S. House of Representatives*, 86th Cong., 1st sess. (4 May 1959) (testimony of George Low).

95. George Low, interview by James M. Grimwood, 7 February 1967, Houston, Texas, George M. Low Papers, 1930–1984.

96. Swanson, *Before This Decade Is Out*, 326.

97. *Hearings before the Committee on Science and Astronautics, U.S. House of Representatives*, 86th Cong., 1st sess. (4 May 1959) (testimony of George Low).

98. *Hearings before the Committee on Science and Astronautics, U.S. House of Representatives*, 86th Cong., 1st sess. (4 May 1959) (testimony of George Low).

99. *Hearings before the Committee on Science and Astronautics, U.S. House of Representatives*, 86th Cong., 1st sess. (4 May 1959) (testimony of George Low).

100. Tex Easley, "Astronauts Find a Friend in Texas Solon Teague," *Abilene Reporter-News*, 15 December 1966.

101. *Hearings before the Committee on Science and Astronautics, U.S. House of Representatives*, 86th Cong., 1st sess. (4 May 1959) (testimony of George Low).

102. United Press International, "Red Space Man Going Up Soon," 10 September 1960.

103. Bob Considine, "Daring Young Men," syndicated newspaper column, 6 February 1961.

104. Considine, "Daring Young Men."

105. Grady, "George Low."

106. Swanson, *Before This Decade Is Out*, 328.

107. Christopher Kraft, interview by the author, 20 December 2017.

108. Logsdon, *Exploring the Unknown*, 389.

109. Logsdon, *Exploring the Unknown*, 389.

110. Logsdon, *Exploring the Unknown*, 389.

111. Logsdon, *Decision to Go to the Moon*, 34–35.

112. Dethloff, *Suddenly, Tomorrow Came*, 28.

113. Dethloff, *Suddenly, Tomorrow Came*, 28.

114. George Low, interview by Robert Sherrod, 28 June 1972.

115. George Low, interview by Robert Sherrod, 28 June 1972.

116. George Low, interview by Robert Sherrod, 28 June 1972.

117. George Low, interview by Robert Sherrod, 7 November 1969.

118. George Low, interview by James Burke, 25 May 1979, BBC, NASA History Office.

119. Jan Armstrong, email to Diane Murphy, 22 October 2011.

120. George Low, memo to the Honorable Olin E. Teague, 10 November 1966, George M. Low Papers, 1930–1984.

121. George Low, memo to the Honorable Olin E. Teague, 10 November 1966.

122. Swanson, *Before This Decade Is Out*, 329.

123. United Press International, "Color Film Shows Shepard Calm during Entire Trip," 11 May 1961.

124. Swanson, *Before This Decade Is Out*, 329.

3. Toward a Worthy Goal

1. George Low, STG interviews, 9 January 1969, 14 January 1969, 4 February 1969, transcript memos, George M. Low Papers, 1930–1984.

2. Wiesner committee, "Report to the President-elect of the Ad Hoc Committee on Space," 10 January 1961, NASA History Office, https://www.hq.nasa.gov/office/pao/History/report61.html.

3. Logsdon, *Decision to Go to the Moon*, 390.

4. James E. Webb, "The Space Age Is Changing Your Life on Earth," World Book Encyclopedia Science Service, 21 May 1961.

5. George Low, STG interviews, 9 January 1969, 14 January 1969, 4 February 1969, transcript memos, George M. Low Papers, 1930–1984.

6. John F. Kennedy, Special Message to the Congress on Urgent National Needs, May 25, 1961, in *Public Papers of the Presidents of the United States: John F. Kennedy*, 403–4.

7. George Low, STG interviews, 9 January 1969, 14 January 1969, 4 February 1969, transcript memos, George M. Low Papers, 1930–1984.

8. Robert Gilruth, "Experts Were Stunned by the Scope of the Mission," *New York Times*, 17 July 1969.

9. George Low, interview by Grimwood, Hacker, and Vorzimmer, 7 February 1967, NASA History Office.

10. George Low, interview by Grimwood, Hacker, and Vorzimmer, 7 February 1967, NASA History Office.

11. George Low, interview by Grimwood, Hacker, and Vorzimmer, 7 February 1967, NASA History Office.

12. "Space Scientists Reveal Moon Rocket Program," *New York Times*, 26 May 1961.

13. United Press International, "Action Now on Man on Moon Shot," 28 May 1961.

14. Alton Blakeslee, "Space Age Need: Brains," Associated Press Science Report, 27 May 1961.

15. Dethloff, *Suddenly, Tomorrow Came*, 31.

16. Logsdon, *Decision to Go to the Moon*, 56.

17. Logsdon, *Decision to Go to the Moon*, 388.

18. Benson and Faherty, *Moonport*, 12.

19. George Low, interview by Grimwood, Hacker, and Vorzimmer, 7 February 1967, NASA History Office.

20. Brooks, Grimwood, and Swenson, *Chariots for Apollo*, 31–37.

21. George Low, interview by Grimwood, Hacker, and Vorzimmer, 7 February 1967, NASA History Office.

22. T. Keith Glennan, tribute speech at George Low's NASA farewell ceremony, 26 May 1976.

23. Logsdon, *Decision to Go to the Moon*, 59.

24. George Low, interview by Grimwood, Hacker, and Vorzimmer, 7 February 1967, NASA History Office.

25. Brooks, Grimwood, and Swenson, *Chariots for Apollo*, 15.

26. Vern Haugland, "$12–15 Billion Cost: Space Probe to Be Vast, NASA Says," Associated Press, 29 July 1960.

27. Swanson, *Before This Decade Is Out*, 313.

28. Logsdon, *Decision to Go to the Moon*, 59.

29. "RPI Graduate Explains Aims of 'Moon Ship,'" *Troy (NY) Times Record*, 30 July 1960.

30. United Press International, "Space Capsule Blasted to Bits," 29 July 1960.

31. Brooks, Grimwood, and Swenson, *Chariots for Apollo*, 16–29.

32. George Low, interview by Addison M. Rothrock, Eugene M. Emme, and Jay Holmes, 1 May 1964, Oral History Project, John F. Kennedy Presidential Library.

33. George Low, interview by Grimwood, Hacker, and Vorzimmer, 7 February 1967, NASA History Office.

34. Brooks, Grimwood, and Swenson, *Chariots for Apollo*, 19–20.

35. George Low, interview by Robert Sherrod, 15 August 1969.

36. George Low, interview by Addison M. Rothrock, Eugene M. Emme, and Jay Holmes, 1 May 1964, Oral History Project, John F. Kennedy Presidential Library.

37. George Low, NASA memo to Abe Silverstein, 17 October 1960, George M. Low Papers, 1930–1984.

38. Brooks, Grimwood, and Swenson, *Chariots for Apollo*, 22.

39. Logsdon, *Decision to Go to the Moon*, 61.

40. George M. Low, Manned Lunar Working Group, *A Plan for a Manned Lunar Landing* (Washington DC: NASA, 7 February 1961), in Logsdon, *Exploring the Unknown*.

41. James Webb, tribute speech at George Low's NASA farewell ceremony, 26 May 1976.

42. Logsdon, *Decision to Go to the Moon*, 90.

43. Seamans, *Project Apollo*, 12.

44. Shepard and Slayton, *Moon Shot*, 214.

45. "In Memoriam: George M. Low," *Roundup*, 20 July 1984, 4.

46. Abe Silverstein, tribute speech at George Low's NASA farewell ceremony, 26 May 1976, recording in author's possession.

47. George Low, interview by Robert Sherrod, 5 July 1972.

48. George Low, interview by Robert Sherrod, 5 July 1972.

49. George Low, interview by Robert Sherrod, 21 June 1972.

50. William Yadley, "D. Brainerd Holmes, a Leader in the Space Race, Dies at 91," *New York Times*, 15 January 2013.

51. Yadley, "D. Brainerd Holmes, a Leader in the Space Race, Dies at 91."

52. George Low, interview by Robert Sherrod, 5 July 1972.

53. George Low, interview by Robert Sherrod, 16 January 1974.

54. George Low, interview by Robert Sherrod, 7 November 1969.

55. George Low, interview by Robert Sherrod, 5 July 1972.

56. George Low, interview by Robert Sherrod, 7 November 1969.

57. George Low, interview by Robert Sherrod, 7 November 1969.

58. George Low, interview by Robert Sherrod, 14 February 1970.

59. Richard S. Johnston, interview by Summer Chick Bergen, 3 November 1998, NASA Johnson Space Center Oral History Project.

60. Ben Evans, "Examinations of Some Kind: The Story of America's First Spacewalk, Part 1," AmericaSpace, 30 May 2015, http://americaspace.com/2015/05/30/examinations-of-some-kind-the-story-of-Americas-first-spacewalk-part-1.

61. *National Observer*, 17 April 1967.

62. John W. Holland, interview by Sandra Johnson, 19 February 2004, NASA Johnson Space Center Oral History Project.

63. George Low, interview by Robert Sherrod, 14 February 1970.

64. "NASA Budgets: US Spending on Space Travel since 1958," *Data Blog, Guardian*, 1 February 2010, https://www.theguardian.com/news/datablog/2010/feb/01/nasa-budgets-us-spending-space-travel.

65. George Low, interview by John Logsdon, 7 July 1970, NASA History Office.

66. George Low, Apollo Notes for Dr. Gilruth, no. 57, 5 July 1967.

67. George Low to Frederick J. Lees, chairman, NASA Inventions and Contributions Board, 21 October 1982, George M. Low Papers, 1930–1984.

68. George Low, interview by Robert Sherrod, 12 August 1970.

69. George Low, interview by Robert Sherrod, 5 July 1972.

70. George Low, interview by Robert Sherrod, 5 July 1972.

71. Furlong, "Biggest Headache in Space."

72. George Low, interview by W. S. Pooler, 7 January 1971, George M. Low Papers, 1930–1984.

73. George Low, interview by Robert Sherrod, 5 July 1972.

4. The Longest Days

1. Furlong, "Biggest Headache in Space."

2. Eric Berger, "The Hell of *Apollo 1*: Pure Oxygen, a Single Spark, and Death in 17 Seconds," *Ars Technica*, 23 January 2017, https://arstechnica.com/science/2017/01/the-hell-of-apollo-1-pure-oxygen-a-single-spark-and-death-in-17-seconds/.

3. Berger, "Hell of Apollo 1."

4. George Low, interview by James Burke, *Project Apollo*, BBC TV, 25 May 1979.

5. George M. Low, "In Developing Space Hardware, Human Judgment Still Counts Most," *New York Times*, 17 July 1969.

6. Murray and Cox, *Apollo*, 198–99.

7. George Low, STG interviews, 9 January 1969, 14 January 1969, 4 February 1969, transcript memos, p. 20, George M. Low Papers, 1930–1984.

8. "Apollo 1 Lessons and Legacies," panel discussion, Johnson Space Center, Houston TX, 24 January 2017, video, 58:12, https://youtu.be/ZxBvAB_ekTQ.

9. Andrew Chaikin, "Apollo's Worst Day," *Air and Space Magazine*, November 2016.

10. Ronald Thompson, "Old Drive Returning to U.S. Space Team after Apollo Deaths," Associated Press, 10 April 1967.

11. George Low, interview by James Burke, *Project Apollo*, BBC TV, 25 May 1979.

12. Keith Wheeler, "Disaster—The Harsh Schoolmaster," *Life*, 26 January 1968, 56.

13. Murray and Cox, *Apollo*, 220.

14. George Low, interview by James Burke, *Project Apollo*, BBC TV, 25 May 1979.

15. Low, "Spaceships," 59.

16. "Nation: The Groundling Who Won," *Time*, 3 January 1969.

17. George Low, interview by Robert Sherrod, 5 July 1972.

18. Jerry Bostick, interview by the author, 10 March 2016.

19. Brian Welch, "*Apollo 8*: Twenty Years Ago, a Bold Maneuver Became Salvation of Troubled Year," *Roundup*, 16 December 1988, 1.

20. "Apollo 1 Lessons and Legacies."

21. Kranz, *Failure Is Not an Option*, 212.

22. "Apollo 1 Lessons and Legacies."

23. Murray and Cox, *Apollo*, 222.

24. George Low, Apollo Notes for Dr. Gilruth, no. 2, 11 April 1967.

25. Gerry Griffin, interview by the author, 9 March 2016.

26. Brian Welch, "In Memoriam, George M. Low, 1926–1984," *Roundup*, 20 July 1984, 4.

27. Christopher Kraft, interview by Rebecca Wright, 23 May 2008, NASA Johnson Space Center Oral History Project.

28. "Moon: Who Made It Possible," *Time*, 18 July 1969.

29. George Abbey, interview by the author, 20 July 2017.

30. George Low, STG interviews, 9 January 1969, 14 January 1969, 4 February 1969, transcript memos, pp. 16–17, George M. Low Papers, 1930–1984.

31. George Low, interview by Robert Sherrod, 5 July 1972.

32. Christopher Kraft, interview by the author, 20 December 2017.

33. Christopher Kraft, interview by Rebecca Wright, 23 May 2008, NASA Johnson Space Center Oral History Project.

34. George Low, STG interviews, 9 January 1969, 14 January 1969, 4 February 1969, transcript memos, pp. 16–17, George M. Low Papers, 1930–1984.

35. Wheeler, "Disaster—The Harsh Schoolmaster," 56.

36. "Space: Fireproofing Apollo," *Time*, 1 September 1967.

37. George Low, interview by James Burke, *Project Apollo*, BBC TV, 25 May 1979.

38. Murray and Cox, *Apollo*, 209.

39. George Low, interview by Robert Sherrod, 5 July 1972.

40. Low, "Spaceships," 75.

41. "Apollo 1 Lessons and Legacies."

42. George Low, STG interviews, 9 January 1969, 14 January 1969, 4 February 1969, transcript memos, pp. 24–25, George M. Low Papers, 1930–1984.

43. George Low, interview by W. S. Pooler, 7 January 1971, George M. Low Papers, 1930–1984.

44. George Low, STG interviews, 9 January 1969, 14 January 1969, 4 February 1969, transcript memos, p. 17, George M. Low Papers, 1930–1984.

45. George Low, interview by W. S. Pooler, 7 January 1971, George M. Low Papers, 1930–1984.

46. George Low, interview by W. S. Pooler, 7 January 1971, George M. Low Papers, 1930–1984.

47. George Low, Apollo Notes for Dr. Gilruth, no. 65, 15 July 1967.

48. George Low, Apollo Notes for Dr. Gilruth, no. 64, 14 July 1967.

49. Christopher Kraft, interview by the author, 20 December 2017.

50. Low, "Spaceships," 69.

51. Low, "Spaceships," 73.

52. T. K. Mattingly, interview by Kevin Rusnak, 22 April 2002, NASA Johnson Space Center Oral History Project.

53. George Abbey, interview by the author, 20 July 2017.

54. Low, introduction, 6.

55. Christopher Kraft, interview by the author, 20 December 2017.

56. Low, "Spaceships," 75.

57. George M. Low, memo to Charles Frick, Apollo spacecraft program office manager, 29 September 1962, George M. Low Papers, 1930–1984.

58. Low, "Spaceships," 75.

59. George Abbey, interview by the author, 20 July 2017.

60. George Abbey, interview by the author, 20 July 2017.

61. Low, "Spaceships," 73.

62. Low, "Spaceships," 69.

63. George Low, Apollo Notes for Dr. Gilruth, no. 31, 18 May 1967.

64. Frank Borman, interview by the author, 10 April 2016.

65. Borman, *Countdown*, 182.

66. George Abbey, interview by the author, 20 July 2017.

67. George Low, Apollo Notes for Dr. Gilruth, no. 1, 10 April 1967.

68. George Low, Apollo Notes for Dr. Gilruth, no. 3, 12 April 1967.

69. George Low, Apollo Notes for Dr. Gilruth, no. 8, 18 April 1967.

70. George Low, Apollo Notes for Dr. Gilruth, no. 143, 29 November 1967.

71. Low, "Spaceships," 61.

72. Welch, "In Memoriam, George M. Low, 1926–1984," 3.

73. George Low, Apollo Notes for Dr. Gilruth, no. 43, 7 June 1967.

74. George Low, Apollo Notes for Dr. Gilruth, no. 38, 1 June 1967.

75. George Low, interview by W. S. Pooler, 7 January 1971, George M. Low Papers, 1930–1984.

76. George Low, interview by James Burke, *Project Apollo*, BBC TV, 25 May 1979.

77. George Low, interview by James Burke, *Project Apollo*, BBC TV, 25 May 1979.

78. "Nation: The Groundling Who Won."

79. George Low, Apollo Notes for Dr. Gilruth, no. 367, 18 October 1968.

80. Nancy Sullivan, interview by the author, 31 May 2016.

81. George Low, Apollo Notes for Dr. Gilruth, no. 81, 21 August 1967.

82. George Low, Apollo Notes for Dr. Gilruth, no. 112, 6 October 1967.

83. Christopher Kraft, interview by the author, 20 December 2017.

84. George Low, interview by Robert Sherrod, 16 January 1974.

85. Low, introduction, 6.

86. O'Toole, "George Low."

87. "Space: Fireproofing Apollo."

88. George M. Low, speech before the subcommittee on energy research and production on nuclear safety after the Three Mile Island nuclear accident, 24 May 1979, George M. Low Papers, 1930–1984.

89. George Low, interview by James Burke, *Project Apollo*, BBC TV, 25 May 1979.

90. George Low, interview by James Burke, *Project Apollo*, BBC TV, 25 May 1979.

91. George Low, STG interviews, 9 January 1969, 14 January 1969, 4 February 1969, transcript memos, p. 18, George M. Low Papers, 1930–1984.

92. George Low, Apollo Notes for Dr. Gilruth, no. 7, 17 April 1967.

93. Grady, "George Low," 48.

94. George Low, commencement speech, University of Florida, Gainesville, 15 June 1969, George M. Low Papers, 1930–1984.

95. Doug Ward, interview by the author, 30 January 2016.

96. George Abbey, interview by the author, 20 July 2017.

97. George Low, Apollo Notes for Dr. Gilruth, no. 76, 31 July 1967.

98. George Abbey, interview by the author, 20 July 2017.

99. Grady, "George Low," 44.

100. George Low, Apollo Notes for Dr. Gilruth, no. 47, 13 June 1967.

101. George Low, Apollo Notes for Dr. Gilruth, no. 69, 22 July 1967.

102. George Low, interview by Robert Sherrod, 16 January 1974.

103. George Low, Apollo Notes for Dr. Gilruth, no. 26, 11 May 1967.

104. George Low, interview by James Burke, *Project Apollo*, BBC TV, 25 May 1979.

105. "Apollo 1 Lessons and Legacies."

106. Low, introduction, 7.

107. Gerry Griffin, interview by the author, 9 March 2016.

108. Low, "Spaceships," 60.

109. Shepard and Slayton, *Moon Shot*, 215.

110. Welch, *"Apollo 8,"* 1.

111. Welch, *"Apollo 8,"* 4.

112. George Low, interview by James Burke, *Project Apollo*, BBC TV, 25 May 1979.

113. George Low, Apollo Notes for Dr. Gilruth, no. 308, 12 July 1968.

114. George Low, Apollo Notes for Dr. Gilruth, no. 308, 12 July 1968.

115. George Low, interview by James Burke, *Project Apollo*, BBC TV, 25 May 1979.

116. George Low, Apollo Notes for Dr. Gilruth, no. 5, 14 April 1967.

117. Gene Kranz, interview by Roy Neal, 19 March 1998, NASA Johnson Space Center Oral History Project.

118. Welch, *"Apollo 8,"* 1.

119. George Low, interview by James Burke, *Project Apollo*, BBC TV, 25 May 1979.

120. George M. Low, memo to the NASA History Office, 29 September 1975, George M. Low Papers, 1930–1984.

121. George Low, special notes for 9 August 1968, George M. Low Papers, 1930–1984; George Low, special notes for *Apollo 8* decision, p. 1, George M. Low Papers, 1930–1984.

122. George Low, special notes for *Apollo 8* decision, p. 4, George M. Low Papers, 1930–1984.

123. George Low, special notes for *Apollo 8* decision, p. 4, George M. Low Papers, 1930–1984.

124. George Low, special notes for *Apollo 8* decision, p. 4, George M. Low Papers, 1930–1984.

125. George Low, special notes for *Apollo 8* decision, p. 4, George M. Low Papers, 1930–1984.

126. George Low, special notes for *Apollo 8* decision, p. 7, George M. Low Papers, 1930–1984.

127. George Low, special notes for *Apollo 8* decision, p. 7, George M. Low Papers, 1930–1984.

128. George Low, special notes for *Apollo 8* decision, p. 9, George M. Low Papers, 1930–1984.

129. George Low, special notes for *Apollo 8* decision, p. 11, George M. Low Papers, 1930–1984.

130. George Low, special notes for *Apollo 8* decision, p. 11, George M. Low Papers, 1930–1984.

131. George Low, interview by James Burke, *Project Apollo*, BBC TV, 25 May 1979.

132. George Low, special notes for *Apollo 8* decision, p. 11, George M. Low Papers, 1930–1984.

133. Christopher Kraft, interview by Rebecca Wright, 23 May 2008, NASA Johnson Space Center Oral History Project.

134. Dill Hunley, George M. Low profile, George M. Low Papers, 1930–1984, ser. 1, box 1, folder 1, #4133.

135. George Low, STG interviews, 9 January 1969, 14 January 1969, 4 February 1969, transcript memos, p. 25, George M. Low Papers, 1930–1984.

136. Welch, "*Apollo 8*," 1.

137. Welch, "*Apollo 8*," 4.

138. Gene Kranz, interview by Roy Neal, 19 March 1998, NASA Johnson Space Center Oral History Project.

139. Welch, "*Apollo 8*," 4.

140. George Low, STG interviews, 9 January 1969, 14 January 1969, 4 February 1969, transcript memos, p. 24, George M. Low Papers, 1930–1984.

141. George Low, commencement speech, University of Florida, Gainesville, 15 June 1969, George M. Low Papers, 1930–1984.

142. George Low, Apollo Notes for Dr. Gilruth, no. 365, 1 October 1968.

143. George W. S. Abbey, Apollo Notes for Dr. Gilruth, no. 370, 23 and 24 October 1968.

144. George Low, Apollo Notes for Dr. Gilruth, no. 379, 7 November 1968.

145. George Low, special notes for 10 and 11 November 1968, p. 1, George M. Low Papers, 1930–1984.

146. George Low, special notes for 10 and 11 November 1968, p. 2, George M. Low Papers, 1930–1984.

147. George Low, special notes for 10 and 11 November 1968, p. 2, George M. Low Papers, 1930–1984.

148. George Low, special notes for 10 and 11 November 1968, pp. 3–4, George M. Low Papers, 1930–1984.

149. George Low, special notes for 10 and 11 November 1968, p. 3, George M. Low Papers, 1930–1984.

150. George Low, special notes for 10 and 11 November 1968, p. 4, George M. Low Papers, 1930–1984.

151. George Low, special notes for 10 and 11 November 1968, p. 5, George M. Low Papers, 1930–1984.

152. George Low, special notes for 10 and 11 November 1968, p. 6, George M. Low Papers, 1930–1984.

153. George Low, special notes for 10 and 11 November 1968, p. 6, George M. Low Papers, 1930–1984.

154. Frank Borman, interview by the author, 10 April 2016.

155. George Low, Apollo Notes for Dr. Gilruth, no. 404, 16 December 1968.

156. George Low, commencement speech, University of Florida, Gainesville, 15 June 1969, George M. Low Papers, 1930–1984.

157. George Low, Apollo Notes for Dr. Gilruth, no. 397, 4 December 1968.

158. George Low, Apollo Notes for Dr. Gilruth, no. 403, 13 December 1968.

159. Zimmerman, *Genesis*, 95.

160. Kranz, *Failure Is Not an Option*, 245.

161. Zimmerman, *Genesis*, 217.

162. "Nation: The Groundling Who Won."

163. George Low, interview by Robert Sherrod, 16 January 1974.

164. George M. Low, speech at the joint meeting of the Capital District Chambers of Commerce, Troy, New York, 19 April 1979, George M. Low Papers, 1930–1984.

165. David Templeton, "He Came from the City, Worked on Apollo Memorials," *Pittsburgh Post-Gazette*, 20 July 2009.

166. George Low, interview by James Burke, *Project Apollo*, BBC TV, 25 May 1979.

167. Templeton, "He Came from the City."

168. George Low, NASA letter to Robert Sherrod, 15 August 1969, George M. Low Papers, 1930–1984.

169. George Low, Apollo Notes for Dr. Gilruth, no. 413, 10 January 1969.

170. Gerry Griffin, interview by the author, 9 March 2016.

171. Kraft, *Flight*, 323.

172. Kraft, *Flight*, 314.

173. George Low, NASA memo to Julian Scheer, 18 March 1969, George M. Low Papers, 1930–1984.

174. Julian Scheer, NASA memo to George Low, 12 March 1969, George M. Low Papers, 1930–1984.

175. George Low, NASA letter to Mrs. Janet Burdick, 3 April 1969, George M. Low Papers, 1930–1984.

176. George Low, NASA memo to Julian Scheer, 18 March 1969, George M. Low Papers, 1930–1984.

177. Kraft, *Flight*, 318.

178. Jerry Bostick, interview by the author, 10 March 2016.

179. George Low, interview, July 1974, NASA special report, NASA Public Affairs, recording in author's possession.

180. George Low, interview, July 1974, NASA special report, NASA Public Affairs, recording in author's possession.

181. George Low, interview by Robert Sherrod, 5 July 1972.

182. Kranz, *Failure Is Not an Option*, 236, 340.

183. George Low, interview by Robert Sherrod, 5 July 1972.

184. George Low, interview by Robert Sherrod, 5 July 1972.

5. Post-Apollo

1. Logsdon, *After Apollo?*, 92.

2. John Noble Wilfred, "Soft Deadline for a Trip to Mars," *New York Times*, 21 September 1969.

3. George Low, Personal Notes, no. 1, 1 January 1970, p. 2.

4. Logsdon, *After Apollo?*, 56.

5. George Low, Personal Notes, no. 1, 1 January 1970, p. 1.

6. Logsdon, *After Apollo?*, 86.

7. Logsdon, *After Apollo?*, 86.

8. *Hearing on the Nomination of George M. Low to Be Deputy Administrator of the National Aeronautics and Space Administration before the Committee on Aeronautical and Space Sciences, United States Senate*, 91st Cong., 1st sess. (25 November 1969) (testimony by George M. Low) (Washington DC: U.S. Government Printing Office, 1969), 5.

9. *Hearing on the Nomination of George M. Low to Be Deputy Administrator of the National Aeronautics and Space Administration*, 6.

10. *Hearing on the Nomination of George M. Low to Be Deputy Administrator of the National Aeronautics and Space Administration*, 5.

11. *Hearing on the Nomination of George M. Low to Be Deputy Administrator of the National Aeronautics and Space Administration*, 5.

12. Heppenheimer, "Winter of Discontent."

13. Logsdon, *After Apollo?*, 62–63.

14. Logsdon, *After Apollo?*, 62–63.

15. George Low, Personal Notes, no. 1, 1 January 1970, p. 2.

16. George Low, Personal Notes, no 1, 1 January 1970, p. 2.

17. George Low, interview by John Logsdon, 7 July 1970, NASA History Office.

18. George Low, Personal Notes, no. 38, 3 January 1971, p. 5.

19. George Low, Personal Notes, no. 17, 4 April 1970, p. 2.

20. William F. Hammond Jr., "The Man and the Moon," *Rensselaer*, December 1994, 8.

21. George Low, interview by John Logsdon, 7 July 1970, NASA History Office.

22. George Low, interview by Robert Sherrod, 14 July 1970.

23. Tom Rees, "NASA Director Predicts Exciting Space Feats in 1970s," *Rocky Mountain News*, 9 March 1971.

24. Rees, "NASA Director Predicts Exciting Space Feats in 1970s."

25. George Low, interview by John Logsdon, 7 July 1970, NASA History Office.

26. "No Domestic Cure-All: NASA Boss Belies Claims," *Huntsville (AL) Times*, 24 November 1970.

27. Bill Anders, interview by the author, 28 June 2016.

28. Bill Anders, interview by the author, 28 June 2016.

29. Bill Anders, interview by the author, 28 June 2016.

30. Bill Anders, interview by the author, 28 June 2016.

31. George Low, "RE: Discussions with Bill Anders," EYES ONLY/SENSITIVE memo to the administrator, 11 November 1971, George M. Low Papers, 1930–1984.

32. Bill Anders, interview by the author, 28 June 2016.

33. "NASA's Needs Listed," *Houston Post*, 5 May 1971.

34. George Low, interview by John Logsdon, 7 July 1970, NASA History Office.

35. George Low, interview by John Logsdon, 30 January 1980, NASA History Office.

36. Gerry Griffin, interview by the author, 9 March 2016.

37. George Low, Personal Notes, no. 58, 14 November 1971.

38. Bill Anders, interview by the author, 28 June 2016.

39. George Low, "RE: Furnishing Information to Congress," NASA memo to all associate and assistant administrators, 30 January 1970, George M. Low Papers, 1930–1984.

40. Gerry Griffin, interview by the author, 9 March 2016.

41. Gerry Griffin, interview by the author, 9 March 2016.

42. George Low, Personal Notes, no. 27, 18 July 1970, p. 4.

43. John Conway, interview by the author, 24 March 2016.

44. George Low, Personal Notes, no. 25, 21 June 1970, p. 9.

45. George Low, Personal Notes, no. 27, 18 July 1970, pp. 3–4.

46. George Low, Personal Notes, no. 27, 18 July 1970, p. 5.

47. George Low, Personal Notes, no. 1, 1 January 1970, p. 7.

48. George Low, Personal Notes, no. 25, 21 July 1970, p. 8.

49. George Low, Personal Notes, no. 25, 21 July 1970, p. 8.

50. George Low, Personal Notes, no. 1, 1 January 1970, p. 8.

51. George Low, Personal Notes, no. 14, 7 March 1970, p. 2.

52. George Low, Personal Notes, no. 40, 22 January 1971, p. 3.

53. George Low, Personal Notes, no. 42, 21 February 1971, p. 5.

54. Klaus P. Heiss, *Mathematica Economic Analysis of the Space Shuttle System*, NASA contract NASW-2081, 31 January 1972, NASA Technical Report Server, https://ntrs.nasa.gov/archive/nasa/casi.ntrs.nasa.gov/19730005253.pdf.

55. Mike Wall, "NASA's Shuttle Program Cost $209 Billion: Was It Worth It?" Space.com, 5 July 2011, https://www.space.com/12166-space-shuttle-program-cost-promises-209-billion.html.

56. Bill Anders, interview by the author, 28 June 2016.

57. George Low, "Productivity in the Space Program" (address to the National Space Club, Washington DC, 17 February 1972), George M. Low Papers, 1930–1984.

58. Bill Anders, interview by the author, 28 June 2016.

59. Tariq Malik, "NASA's Space Shuttle by the Numbers: 30 Years of a Space-flight Icon," Space.com, 21 July 2011, https://www.space.com/12376-nasa-space -shuttle-program-facts-statistics.html.

60. George Low, keynote speech, symposium of the National Security Industrial Association and the Armed Forces Management Association, Washington DC, September 1972, George M. Low Papers, 1930–1984.

61. Gene Kranz, interview by Roy Neal, 28 April 1999, NASA Johnson Space Center Oral History Project.

62. Bill Anders, interview by the author, 28 June 2016.

63. Gene Kranz, interview by Roy Neal, 28 April 1999, NASA Johnson Space Center Oral History Project.

64. Richard Nixon, "President Nixon's 1972 Announcement of the Space Shut-tle," 5 January 1972, NASA, last updated 30 March 2009, https://history.nasa .gov/printFriendly/stsnixon.htm.

65. Richard Nixon, White House letter to George M. Low, 19 May 1972, George M. Low Papers, 1930–1984.

66. Richard D. Lyons, "Senate Votes Aid to Space Shuttle," *New York Times*, 12 May 1972.

67. Roger Launius, interview by the author, 21 July 2016.

68. George Low, Personal Notes, no. 43, 7 March 1971, p. 1.

69. George Low, Personal Notes, no. 44, 21 March 1971, p. 1.

70. George Low, interview by John Logsdon, 30 January 1980, NASA History Office.

71. George Low, NASA letter to Miss Marsha Crandall, 2 April 1972, George M. Low Papers, 1930–1984.

72. George Low, Personal Notes, no. 13, 28 February 1970, p. 3.

73. George Low, Personal Notes, no. 42, 21 February 1971, p. 1.

74. George Low, Personal Notes, no 36, 28 November 1970, p. 7.

75. Al Alibrando, interview by Robert Sherrod, 8 April 1974; see also Whye, "Heroes, Not of Their Own Accord," 356.

76. George Low, Personal Notes, no. 44, 10 April 1971, p. 4.

77. Paul Sawyer, memo and report for George Low on NASA Public Affairs, 15 April 1971, pp. 1–3, George M. Low Papers, 1930–1984.

78. George Low, Personal Notes, no. 44, 10 April 1971, p. 3.

79. Jim Fletcher, "RE: NASA Spin-Offs," NASA memorandum for George Low, 4 April 1974, George M. Low Papers, 1930–1984.

80. Doug Ward, interview by the author, 30 January 2016.

81. Doug Ward, interview by the author, 30 January 2016.

82. Dudley G. McConnell to George Low, 18 May 1976, George M. Low Papers, 1930–1984.

83. United Press International, "NASA Discrimination Scored," 29 October 1973.

84. "Killing the Messenger," *Florida Today*, 3 November 1973.

85. George Low, Personal Notes, no. 113, 3 February 1974, p. 2.

86. Carl Hiaasen, "Fired, Hired, Still Speaking," *Today*, 21 August 1974.

87. George Low, Personal Notes, no. 113, 3 February 1974, pp. 2–4.

88. Hiaasen, "Fired, Hired, Still Speaking."

89. Harriett G. Jenkins, interview by Jennifer Ross-Nazzal, 5 August 2011, NASA Johnson Space Center Oral History Project.

90. Harriett G. Jenkins, interview by Jennifer Ross-Nazzal, 5 August 2011, NASA Johnson Space Center Oral History Project.

91. Harriett G. Jenkins, interview by Jennifer Ross-Nazzal, 5 August 2011, NASA Johnson Space Center Oral History Project.

92. Sarah Loff, "1978 Astronaut Class," NASA History, last updated 6 August 2017, https://www.nasa.gov/image-feature/1978-astronaut-class.

93. George Low, Personal Notes, no. 19, 25 April 1970, p. 3.

94. George Low, Personal Notes, no. 25, 21 June 1970, p. 1.

95. George Low, Personal Notes, no. 27, 18 June 1970, p. 1.

96. George Low, Personal Notes, no. 42, 21 February 1971, p. 1.

97. George Low, Personal Notes, no. 52, 15 August 1971, p. 1.

98. George Low, Personal Notes, no. 56, 17 October 1971, p. 2.

99. George Low, interview by Robert Sherrod, 7 September 1972.

100. George Low, Personal Notes, no. 62, 15 January 1972, p. 7.

101. George Low, Personal Notes, no. 71, 3 June 1972, p. 4.

102. George Low, addendum to Personal Notes, no. 74, 18 July 1972, pp. 1–5.

103. George Low, Personal Notes, no. 73, 1 July 1972, p. 10.

104. George Low, NASA letters, 26 July 1972; these letters went out to Mr. Thomas J. Fennaghty, to Mrs. Connie O'Shaughnessy, and to others.

105. George Low, Personal Notes, no. 75, 29 July 1972, p. 2.

106. George Low, Personal Notes, no. 76, 19 August 1972, pp. 1–4.

107. George Low, Personal Notes, no. 76, 19 August 1972, pp. 1–4.

108. George Low, Personal Notes, no. 77, 16 September 1972, p. 1.

109. George Low, Personal Notes, no. 77, 16 September 1972, p. 1.

110. Jim Fletcher, "Autographs," NASA memo to Officials in Charge of Headquarters Program and Staff Offices and Field Installations, 13 November 1972, George M. Low Papers, 1930–1984.

111. George Low, memorandum, 30 October 1972, George M. Low Papers, 1930–1984.

112. George Low, Personal Notes, no. 78, 30 September 1972, p. 2.

113. Paul Sawyer, interview by Robert Sherrod, 2 November 1972, George M. Low Papers, 1930–1984.

114. "Deputy Space Director Outlines Benefits Due Mankind in the Future," *Huntsville (AL) Times*, 6 April 1973.

115. George Low, interview by Bob Thomas, *Today*, NBC, 27 June 1973.

116. George Low, interview by Bob Thomas, *Today*, 27 June 1973.

117. George Low, interview by Bob Thomas, *Today*, 27 June 1973.

118. Phil Smith, "Space Picture Bright, as Painted by Dr. Low," *Huntsville (AL) News*, 6 April 1973.

119. Smith, "Space Picture Bright."

120. Smith, "Space Picture Bright."

121. "Kennedy Proposes Joint Mission to the Moon," History.com, last updated 27 February 2019, https://www.history.com/this-day-in-history/kennedy-proposes-joint-mission-to-the-moon.

122. George Low, interview by Edward Clinton Ezell, 30 April 1975, NASA History Office.

123. George Low, "Meeting with the President," memorandum, 5 January 1972, NASA History Office.

124. George Low, Personal Notes, no. 34, 7 November 1970, p. 4.

125. Ezell and Ezell, *Partnership*, 25.

126. George Low, interview by Edward Clinton Ezell, 30 April 1975, NASA History Office.

127. George Low, interview by Edward Clinton Ezell, 30 April 1975, NASA History Office.

128. George Low, Personal Notes, no. 40, 22 January 1971, p. 3.

129. George Low, interview by Edward Clinton Ezell, 30 April 1975, NASA History Office.

130. George Low, interview by Edward Clinton Ezell, 30 April 1975, NASA History Office.

131. George Low, interview by Edward Clinton Ezell, 30 April 1975, NASA History Office.

132. George Low, interview by Edward Clinton Ezell, 30 April 1975, NASA History Office.

133. Glynn Lunney, interview by the author, 28 April 2016.

134. George Low, interview by Edward Clinton Ezell, 30 April 1975, NASA History Office.

135. George Low, Personal Notes, no. 70, 21 May 1972, p. 6.

136. "Science: Mission Misfire," *Time*, 21 April 1975.

137. "Space: Hands All Round and Four for Dinner," *Time*, 28 July 1975.

138. Ezell and Ezell, *Partnership*, 183.

139. Ezell and Ezell, *Partnership*, 131.

140. "U.S. Space Team at Soviet Center," *New York Times*, 19 October 1973.

141. "U.S. Space Team at Soviet Center."

142. John Noble Wilford, "U.S.-Soviet Meeting in Space: Joint Mission Ready to Go," *New York Times*, 14 July 1975.

143. George Low, Personal Notes, no. 149, 26 July 1975, pp. 3–4.

144. Associated Press, "Chapter in U.S. Space Program Ends: Last Apollo Comes Home," 24 July 1975.

145. George Low, Personal Notes, no. 149, 26 July 1975, pp. 3–4.

6. Giving Back to the Future

1. Ritchie, *Astronautics and Aeronautics, 1976*, 1.

2. George Low, Personal Notes, no. 145, 31 May 1975, p. 3.

3. George Low, Personal Notes, no. 143, 4 May 1975, p. 7.

4. Ritchie, *Astronautics and Aeronautics, 1976*, 50.

5. George Low, "Summary of Considerations—FAA," Handwritten Notes, 22 April 1975, George M. Low Papers, 1930–1984.

6. George Low, "Summary of Considerations—FAA," Handwritten Notes, 22 April 1975, George M. Low Papers, 1930–1984.

7. George Low, "Summary of Considerations—FAA," Handwritten Notes, 22 April 1975, George M. Low Papers, 1930–1984.

8. George Low, "Summary of Considerations—FAA," Handwritten Notes, 22 April 1975, George M. Low Papers, 1930–1984.

9. George Low, "Summary of Considerations—FAA," Handwritten Notes, 22 April 1975, George M. Low Papers, 1930–1984.

10. William N. Walker, "RE: FAA Administrator," memorandum for the president, via Donald Rumsfeld, 28 April 1975, George M. Low Papers, 1930–1984.

11. George Low, second epilogue to special notes on consideration as FAA administrator, 26 March 1976, George M. Low Papers, 1930–1984.

12. George Low to Mr. Elmer R. Davis, 9 May 1975, George M. Low Papers, 1930–1984.

13. Mark Low, email correspondence with the author, 21 September 2016.

14. Rusty Schweickart, interview by the author, 27 May 2016.

15. Rusty Schweickart, interview by the author, 27 May 2016.

16. Rusty Schweickart, interview by the author, 27 May 2016.

17. "Jacques Cousteau at NASA Headquarters," in *Space Colonies: A CoEvolution Book*, ed. Stewart Brand (1977), https://space.nss.org/settlement/nasa/CoEvolutionBook/JCOUST.HTML.

18. Rusty Schweickart, interview by the author, 27 May 2016.

19. George Low, handwritten pro/con notes on TRENCOR vs. RPI, March 1976, George M. Low Papers, 1930–1984.

20. *Roundup*, 26 March 1976.

21. George Low, Personal Notes, no. 167, 4 June 1976, p. 3.

22. Joseph P. Allen, interview by Jennifer Ross-Nazzal, 16 March 2004, NASA Johnson Space Center Oral History Project.

23. McCurdy, *Inside NASA*, 88–89.

24. George Low, handwritten center directors meeting notes, 21 April 1976, George M. Low Papers, 1930–1984.

25. George Low, remarks at farewell retirement dinner, 26 May 1976, George M. Low Papers, 1930–1984.

26. Gary Malone, "NASA's George Low Named RPI President," *Troy (NY) Times Record*, 20 March 1976.

27. Editorial, "Opinion: President Low," *Troy (NY) Times Record*, 22 March 1976.

28. Joe Cooley, "We Love It," *Troy (NY) Times Record*, 27 March 1976.

29. Joe Phillips, editorial, *Polytechnic*, 8 April 1976, 1.

30. Joe Phillips, editorial, *Polytechnic*, 8 April 1976, 1.

31. Joe Phillips, editorial, *Polytechnic*, 8 April 1976, 1.

32. Joe Phillips, editorial, *Polytechnic*, 8 April 1976, 1.

33. Joe Phillips, editorial, *Polytechnic*, 8 April 1976, 1.

34. Editorial, *Polytechnic*, 8 April 1976, 8.

35. Mark Tortorici, "President George M. Low: The Times behind the Man," *Polytechnic*, 13 October 1976, 11.

36. George Low, Personal Notes, no. 166, 21 May 1976, p. 6.

37. George Low, inaugural address, Rensselaer Polytechnic Institute, Troy, New York, 8 October 1976, George M. Low Papers, 1930–1984.

38. George Low, inaugural address, Rensselaer Polytechnic Institute, Troy, New York, 8 October 1976, George M. Low Papers, 1930–1984.

39. Tortorici, "President George M. Low," 11.

40. Editorial, *Polytechnic*, 13 October 1976, 5.

41. George Low, "Rensselaer in 1977: Address to Faculty and Staff," 29 March 1977, George M. Low Papers, 1930–1984.

42. Low, "Rensselaer in 1977."

43. Dorothy Reynolds, interview by the author, 18 May 2016.

44. Wikipedia, s.v. "Frederick Terman," last modified 26 December 2018, 14:07, https://en.wikipedia.org/wiki/Frederick_Terman.

45. Gene Bylinsky, "California's Great Breeding Ground for Industry," *Fortune Magazine*, June 1974.

46. Robert M. Whitaker, "As We Mourn President Low at Rensselaer," *Rensselaer*, Summer 1984, 3.

47. George Low, handwritten speech notes, 24 September 1981, George M. Low Papers, 1930–1984.

48. Hammond, "Man and the Moon."

49. Hammond, "Man and the Moon."

50. Grady, "George Low."

51. Dorothy Reynolds, interview by the author, 18 May 2016.

52. George Low, "Productivity, Professionalism, and the Public Interest" (keynote presentation, Roy V. Wright Lecture, American Society of Mechanical Engineers, Chicago, Illinois, 19 November 1980), George M. Low Papers, 1930–1984.

53. Low, "Productivity, Professionalism, and the Public Interest."

54. Low, "Productivity, Professionalism, and the Public Interest."

55. United Press International, "Panel to Study Aircraft Designs," 21 December 1979.

56. George Low to D. Philip Handler, Chairman, National Research Council, 24 June 1980, George M. Low Papers, 1930–1984.

57. National Research Council, *Improving Aircraft Safety.*

58. Associated Press, "Better Leaders and Engineers Urged for FAA," 27 June 1980.

59. Associated Press, "Better Leaders and Engineers Urged for FAA."

60. Hansen, McAndrews, and Berkeley, *History of Aviation Safety Oversight in the United States*, 25.

61. "Voyager 1 Flies near Saturn," History.com, last updated 21 August 2018, https://www.history.com/this-day-in-history/voyager-i-flies-near-saturn.

62. Don Kirkman, "Twilight in Space: America Has Begun to Turn Its Back on the Stars," Scripps-Howard News Service, 15 November 1980.

63. Kirkman, "Twilight in Space."

64. James M. Odato, "Mom Beaming as Son Joins Reagan's Staff," *Press and Sun-Bulletin* (Binghamton NY), 19 November 1980.

65. Rick Barry, "Noncommittal but 'Pro-Space' GOP Keeps NASA Fingers Crossed," *Tampa (FL) Tribune-Times*, 14 December 1980.

66. George M. Low, Team Leader, NASA Transition Team, memo to Mr. Richard Fairbanks, Director, Transition Resources and Development Group, 19 December 1980, NASA Historical Reference Collection, NASA History Office, NASA Headquarters, Washington DC, https://www.hq.nasa.gov/office/pao/History/low80.html.

67. George M. Low, *Report of the Transition Team, National Aeronautics and Space Administration* (Washington DC, 1980), https://www.hq.nasa.gov/office/pao/History/low80.html.

68. John M. Logsdon, "National Leadership and Presidential Power," in Launius and McCurdy, *Spaceflight and the Myth of Presidential Leadership*, 213.

69. Low, *A Plan for a Manned Lunar Landing*, 3.

70. Jack Welch, interview by the author, 17 October 2016.

71. George Low, "The Human Side of Quality" (address to General Electric Company Corporate Officers Meeting, Phoenix, Arizona, 6 October 1981), George M. Low Papers, 1930–1984.

72. Mark Low, email correspondence with the author, 21 September 2016.

73. George Abbey, interview by the author, 20 July 2017.

74. Richard A. D'Errico, "The Legacy of George Low," *Albany Business Review*, 1 December 2003.

75. John Noble Wilford, "George M. Low Is Dead at 58; Headed Apollo Space Project," *New York Times*, 18 July 1984.

76. Ronald Reagan, "Remarks at White House Ceremony Marking the Fifteenth Anniversary of the *Apollo 11* Lunar Landing," 20 July 1984, Ronald Reagan Presidential Library and Museum, https://www.reaganlibrary.gov/research/speeches/72084d.

77. 130 Cong. Rec. S20,496–97 (daily ed. 23 July 1984) (statement of Senator Moynihan).

Epilogue

1. Leonard David, "Mars Rovers Are in Good Shape for Winter," NBCNews.com, last updated April 25, 2006, http://www.nbcnews.com/id/12480225/ns/technology_and_science-space/t/mars-rovers-are-good-shape-winter/#.XLXSuOhKiHt.

2. Tracey Leibach, "Out of This World Connections," *Rensselaer*, Fall 2006, online edition, https://www.rpi.edu/magazine/fall2006/onelastthing.html.

3. McCurdy, *Inside NASA*, 25.

4. Jack Welch, *General Electric News*, 20 July 1984.

5. Frank Borman, interview by the author, 10 April 2016.

6. Marvin Miles, "Space Program Value Lauded," *Los Angeles Times*, 3 April 1975.

7. Carlos Byars, "Low Is Still Concerned with Safety in the Skies," *Houston Chronicle*, 11 February 1980.

8. NASA, "NASA Renames Excellence Award for George M. Low," press release, 22 October 1990.

9. D'Errico, "Legacy of George Low."

Bibliography

Archival Sources

Aviation Week and Technology magazine, online archive of past issues.

George M. Low Papers, 1930–1984. Folsom Library, Institute Archives and Special Collections, Rensselaer Polytechnic Institute, Troy NY. In addition to the voluminous memos, correspondence, notes, diary entries, calendars, and collected papers that compose the massive George M. Low Archive at Rensselaer Polytechnic Institute, the following three key categories of material numbering in the hundreds of pages written by Low himself deserve to be singled out. Keenly aware of the historic nature of his work, Low kept detailed, daily notes of his activities and observations during some of the most historic events of his career. They are an invaluable source of raw, direct, in-the-moment thoughts and actions of George Low from early 1967 until his final day at NASA in 1976. As Low himself writes as an introduction to his Personal Notes: "This set of notes constitutes my personal record of my activities as Deputy Administrator of NASA. The notes will be strictly personal, will give my own often frank views about people and events and not intended to represent the official record of events. The principal purpose is to assist me in recollecting these events at some future date should this be necessary."

    Apollo Notes for Dr. Gilruth, nos. 1–532, April 1967 to November 1969.

    Notes for the Record, February 1967 to November 1970.

    Personal Notes, nos. 1–167, January 1970 to June 1976.

Glenn History Collection. NASA John H. Glenn Research Center, Cleveland OH.

John F. Kennedy Presidential Library, Textual Archives, Boston MA.

NASA History Office, NASA Headquarters, Washington DC, including the extensive collection of journalist Robert Sherrod's interviews with George M. Low.

NASA Johnson Space Center Oral History Project, JSC History Office, Houston TX.

Newspapers.com, online archive of newspapers from 1700s to 2000s.

Time magazine, online archive of past issues.

Published Sources

Arrighi, Robert S. *Bringing the Future within Reach: Celebrating 75 Years of the NASA John H. Glenn Research Center.* Cleveland OH: Glenn Research Center, NASA, 2016.

Benson, Charles D., and William Barnaby Faherty. *Moonport: A History of Apollo Launch Facilities and Operations.* Washington DC: Scientific and Technical Information Branch, NASA, 1978.

Bilstein, Roger E. *Orders of Magnitude: A History of the NACA and NASA, 1915–1990.* NASA History Series, NASA SP-4406. Washington DC: Scientific and Technical Information Division, NASA, 1989.

———. *Stages to Saturn: A Technological History of the Apollo/Saturn Launch Vehicles.* Washington DC: Scientific and Technical Information Branch, NASA, 1980.

Bizony, Piers. *The Man Who Ran the Moon: James E. Webb, NASA, and the Secret History of Project Apollo.* New York: Thunder's Mouth Press, 2006.

Borman, Frank. *Countdown: An Autobiography.* With Robert J. Serling. New York: William Morrow, 1988.

Brooks, Courtney G., James M. Grimwood, and Loyd S. Swenson Jr. *Chariots for Apollo: A History of Manned Lunar Spacecraft.* Washington DC: Scientific and Technical Information Branch, NASA, 1979.

Brook-Shepherd, Gordon. *Anschluss: The Rape of Austria.* London: Macmillian, 1963.

Burrows, William E. *This New Ocean: The Story of the First Space Age.* New York: Random House, 1998.

Byrnes, Mark E. *Politics and Space: Image Making by NASA.* Westport: Praeger, 1994.

Cassutt, Michael. *The Astronaut Maker: How One Mysterious Engineer Ran Human Spaceflight for a Generation.* Chicago: Chicago Review Press, 2018.

Chaikin, Andrew. *A Man on the Moon: The Voyages of the Apollo Astronauts.* New York: Penguin Books, 1998.

Ciccone, Mary E. "Personal and Political Advocacy: Displaced Persons and the Diplomacy of International Relief, 1938–1948." Thesis, Seton Hall University, South Orange NJ, 2011. https://scholarship.shu.edu/cgi/viewcontent.cgi?article=1197&context=theses.

Crossfield, Albert Scott, and Clay Blair Jr. *Always Another Dawn: The Story of a Rocket Test Pilot.* Cleveland OH: World Publishing Company, 1960.

Cunningham, Walter. *The All-American Boys.* New York: iBooks, 2003.

Daniels, Roger. *Guarding the Golden Door: American Immigration Policy and Immigrants Since 1882.* New York: Hill and Wang, 2005.

Dawson, Virginia P. *Engines and Innovation: Lewis Laboratory and American Propulsion Technology.* NASA History Series, NASA SP-4306. Washington DC: Scientific and Technical Information Division, NASA, 1991.

DeGroot, Gerard J. *Dark Side of the Moon: The Magnificent Madness of the American Lunar Quest*. New York: New York University Press, 2006.

Dethloff, Henry C. *Suddenly, Tomorrow Came: A History of the Johnson Space Center*. NASA History Series, NASA SP-4307. Houston TX: Lyndon B. Johnson Space Center, 1993.

Dick, Steven J., and Roger D. Launius. *Societal Impact of Spaceflight*. Washington DC: Office of External Relations, History Division, NASA, 2007.

Dickson, Paul. *Sputnik: The Shock of the Century*. New York: Walker Books, 2011.

Ezell, Edward Clinton, and Linda Neuman Ezell. *The Partnership: A History of the Apollo-Soyuz Test Project*. NASA History Series, NASA SP-4209. Washington DC: Scientific and Technical Information Office, NASA, 1978.

Fenrich, Eric Brett. "The Color of NASA: Racial Inclusion in the American Space Program, 1957–1978." PhD diss., University of California, Santa Barbara, 2015.

Foster, Amy E. *Integrating Women into the Astronaut Corps: Politics and Logistics at NASA, 1972–2004*. Baltimore MD: Johns Hopkins University Press, 2011.

French, Francis, and Colin Burgess. *Into That Silent Sea: Trailblazers of the Space Era, 1961–1965*. Lincoln: University of Nebraska Press, 2007.

Gilruth, Robert R. "I Believe We Should Go to the Moon." In *Apollo Expeditions to the Moon*, edited by Edgar M. Cortright. Washington DC: Scientific and Technical Information Office, NASA, 1975.

Grathwol, Robert P., and Donita M. Moorhus. *Building for Peace: U.S. Army Engineers in Europe, 1945–1991*. CMH pub. 45-1-1. Washington DC: Center for Military History and Corps of Engineers, U.S. Army, 2005. https://history.army.mil/html/books/045/45-1-1/CMH_Pub_45-1-1.pdf.

Hansen, James R. *First Man: The Life of Neil Armstrong*. New York: Simon and Schuster, 2005.

———. *Spaceflight Revolution: NASA Langley Research Center from Sputnik to Apollo*. Washington DC: NASA, 1995.

Hansen, Mark, Carolyn McAndrews, and Emily Berkeley. *History of Aviation Safety Oversight in the United States*. DOT/FAA/AR-08/39. Washington DC: U.S. Department of Transportation, Federal Aviation Administration, July 2008.

Harford, James. *Korolev: How One Man Masterminded the Soviet Drive to Beat America to the Moon*. New York: John Wiley and Sons, 1997.

Harris, Gordon L. *The Kennedy Space Center Story*. Merritt Island FL: Public Affairs Office, John F. Kennedy Space Center, NASA, 1974.

Harris, Ruth Bates. *Harlem Princess: The Story of Harry Delaney's Daughter*. New York: Vantage Press, 1991.

Heppenheimer, T. A. "Winter of Discontent." Chap. 4 in *The Space Shuttle Decision*.

NASA History Series, NASA SP-4221. Washington DC: NASA History Office, 1999. http://space.nss.org/the-space-shuttle-decision-chapter-4/.

Jenkins, Dennis R. *Space Shuttle: The History of the National Space Transportation System; The First 100 Missions.* Stillwater: Voyageur Press, 2002.

Kauffman, James L. *Selling Outer Space: Kennedy, the Media, and Funding for Project Apollo, 1961–1963.* Tuscaloosa: University of Alabama Press, 1994.

Kennick, Sylvia A. *Guide to the George M. Low Papers, 1930–1984.* Troy NY: Folsom Library, Institute Archives and Special Collections, Rensselaer Polytechnic Institute, 1988.

Kraft, Christopher C. *Flight: My Life in Mission Control.* New York: Dutton, 2001.

Kranz, Gene. *Failure Is Not an Option: Mission Control from Mercury to Apollo 13 and Beyond.* New York: Simon and Schuster, 2009.

Lambright, W. Henry. *Powering Apollo: James E Webb of NASA.* Baltimore: John Hopkins University Press, 1995.

Launius, Roger D., and Howard E. McCurdy, eds. *Spaceflight and the Myth of Presidential Leadership.* Urbana: University of Illinois Press, 1997.

Layton, Edwin T., Jr., and John H. Lienhard, eds. *History of Heat Transfer: Essays in Honor of the 50th Anniversary of the ASME Heat Transfer Division.* New York: Society of Mechanical Engineers, 1988.

Link, Mae Mills. *Space Medicine in Project Mercury.* NASA History Series, NASA SP-4003. Washington DC: Scientific and Technical Information Division, NASA, 1965.

Logsdon, John M. *After Apollo? Richard Nixon and the American Space Program.* New York: Palgrave Macmillan, 2015.

——— . *The Decision to Go to the Moon: Project Apollo and the National Interest.* Chicago: University of Chicago Press, 1970.

——— , ed. *Exploring the Unknown: Selected Documents in the History of the U.S. Civil Space Program.* Vol. 7, *Human Spaceflight: Projects Mercury, Gemini, and Apollo.* With Roger D. Launius. NASA History Series, NASA SP-4407. Washington DC: Scientific and Technical Information Office, NASA, 2008.

——— . *John F. Kennedy and the Race to the Moon.* New York: Palgrave Macmillan, 2010.

Lovell, James A., and Jeffrey Klugger. *Lost Moon: The Perilous Voyage of* Apollo 13. New York: Houghton Mifflin, 1994.

Low, George M. Introduction to *What Made Apollo a Success?* NASA SP-287. Washington DC: Scientific and Technical Information Office, NASA, 1971.

——— . "Project Mercury: A Pioneering Manned Flight into Space." *Spaceflight* 3, no. 5 (September 1961).

——— . "The Spaceships: Building the First Craft to Cross an Ocean of Space." In

Apollo Expeditions to the Moon, edited by Edgar M. Cortright. Washington DC: Scientific and Technical Information Office, NASA, 1975.

Lunney, Glynn. *Highways into Space*. Self-published, 2014.

Makemson, Harlen. *Media, NASA, and America's Quest for the Moon*. New York: Peter Lang Publishing, 2009.

"Man behind Apollo—George Low." *Rensselaer Review* 6, no. 1 (March 1969).

McCurdy, Howard E. *Inside NASA: High Technology and Organizational Change in the U.S. Space Program*. Baltimore MD: John Hopkins University Press, 1993.

———. *Space and the American Imagination*. 2nd ed. Baltimore MD: Johns Hopkins University Press, 2011.

McDougall, Walter A. . . . *The Heavens and the Earth: A Political History of the Space Age*. Baltimore MD: Johns Hopkins University Press, 1997.

McQuiad, Kim. "Racism, Sexism, and Space Ventures: Civil Rights at NASA in the Nixon Era and Beyond." In *Societal Impact of Spaceflight*, edited by Steven J. Dick and Roger D. Launius. Washington DC: Office of External Relations, History Division, NASA, 2007.

Mersch, C. L. *The Apostles of Apollo: The Journey of the Bible to the Moon and the Untold Stories of America's Race into Space*. Bloomington: iUniverse, 2010.

Michaelson, S. M. "The Tri-Service Program—A Tribute to George M. Knauf, USAF (MC)." *IEEE Transactions on Microwave Theory and Techniques* 19, no. 2 (1971): 131–46.

Murray, Charles, and Catherine Bly Cox. *Apollo: The Race to the Moon*. Burkittsville MD: South Mountain Books, 1989.

National Research Council. *Improving Aircraft Safety*. Washington DC: National Academies Press, 1980.

Nelson, Craig. *Rocket Men: The Epic Story of the First Men on the Moon*. New York: Penguin Books, 2009.

Neufeld, Michael J. *Von Braun: Dream of Space, Engineer of War*. New York: Vintage Books, 2007.

Nowak, Gerhard. "Jüdische Unternehmer im österreichisch-slowakischen Grenzgebeit Angern an der March-Záhorská Ves im 20. Jahrhundert." Master's thesis, University of Vienna, Austria, 2012. http://othes.univie.ac.at/27045/1/2012 -12-31_0308265.pdf.

Phillips, Samuel C. "The Shakedown Cruises." In *Apollo Expeditions to the Moon*, edited by Edgar M. Cortright. Washington DC: Scientific and Technical Information Office, NASA, 1975.

Poole, Robert. *Earthrise: How Man First Saw the Earth*. New Haven CT: Yale University Press, 2008.

Public Papers of the Presidents of the United States: John F. Kennedy. Washington DC: U.S. Government Printing Office, 1962.

Pyle, Rod. *Innovation the NASA Way: Harnessing the Power of Your Organization for Breakthrough Success*. New York: McGraw-Hill, 2014.

Radosh, Ronald, and Joyce Milton. *The Rosenberg File*. 2nd ed. New Haven CT: Yale University Press, 1997.

Ritchie, Eleanor H. *Astronautics and Aeronautics, 1976: A Chronology*. NASA History Series, NASA SP-4021. Washington DC: Scientific and Technical Information Branch, NASA, 1984.

Robb, Walter. *Taking Risks: Getting Ahead in Business and Life*. Waukesha WI: Meadow Brook Farm, 2014.

Roosevelt, Eleanor. *On My Own*. New York: Harper and Brothers, 1958.

Seamans, Robert C., Jr. *Project Apollo: The Tough Decisions*. NASA Monographs in Aerospace History, no. 37. Washington DC: NASA Office of External Relations, History Division, 2005.

Shayler, David J. *Disasters and Accidents in Manned Spaceflight*. Chichester, UK: Springer-Praxis, 2000.

Shepard, Alan, and Deke Slyaton. *Moon Shot: The Inside Story of America's Race to the Moon*. With Jay Barbree and Howard Benedict. Nashville TN: Turner Publishing, 1994.

Stafford, Thomas P. *We Have Capture: Tom Stafford and the Space Race*. With Michael Cassutt. Washington DC: Smithsonian Books, 2002.

Steven-Boniecki, Dwight. *Live TV from the Moon*. Burlington ON: Apogee Books, 2010.

Swanson, Glen E., ed. *Before This Decade Is Out: Personal Reflections on the Apollo Program*. Mineola NY: Dover, 2012.

Swenson, Loyd S., Jr., James M. Grimwood, and Charles C. Alexander. *This New Ocean: A History of Project Mercury*. NASA History Series, NASA SP-4201. Washington DC: Scientific and Technical Information Division, NASA, 1998.

Thomas, Lewis. *Late Night Thoughts on Listening to Mahler's Ninth Symphony*. New York: Penguin Books, 1995.

von Ehrenfried, Manfred "Dutch." *The Birth of NASA: The Work of the Space Task Group, America's First True Space Pioneers*. Chichester, UK: Springer-Praxis, 2016.

Ward, Bob. *Dr. Space: The Life of Wernher von Braun*. Annapolis MD: Naval Institute Press, 2005.

Watkins, Billy. *Apollo Moon Missions: The Unsung Heroes*. Westport CT: Praeger, 2006.

Welzig, Werner, Hanno Biber, and Claudia Resch. *Anschluss: March–April 1938 in Österreich*. Vienna, Austria: Verlag der Österreichischen Akaedemie der Wissenschaften, 2010.

Whitehouse, David. *One Small Step: The Inside Story of Space Exploration*. London: Quercus, 2009.

Whye, Perry Michael. "Heroes, Not of Their Own Accord: An Examination of the Publicity Concerning the United States Astronauts from 1959 to 1972." Master's thesis, Iowa State University, 1977. https://lib.dr.iastate.edu/rtd/16655/.

Wolfe, Tom. *The Right Stuff*. New York: Picador, 1979.

Worden, Alfred. *Falling to Earth: An* Apollo 15 *Astronaut's Journey to the Moon*. With Francis French. Washington DC: Smithsonian Books, 2011.

Zimmerman, Robert. *Genesis: The Story of* Apollo 8, *the First Manned Flight to Another World*. New York: Four Walls Eight Windows, 1998.

Index

Go, Flight! The Unsung Heroes of Mission Control, 1965–1992
Rick Houston and Milt Heflin
Foreword by John Aaron

Infinity Beckoned: Adventuring Through the Inner Solar System, 1969–1989
Jay Gallentine
Foreword by Bobak Ferdowsi

Fallen Astronauts: Heroes Who Died Reaching for the Moon, Revised Edition
Colin Burgess and Kate Doolan with Bert Vis
Foreword by Eugene A. Cernan

Apollo Pilot: The Memoir of Astronaut Donn Eisele
Donn Eisele
Edited and with a foreword by Francis French
Afterword by Susie Eisele Black

Outposts on the Frontier: A Fifty-Year History of Space Stations
Jay Chladek
Foreword by Clayton C. Anderson

Come Fly with Us: NASA's Payload Specialist Program
Melvin Croft and John Youskauskas
Foreword by Don Thomas

Shattered Dreams: The Lost and Canceled Space Missions
Colin Burgess
Foreword by Don Thomas

The Ultimate Engineer: The Remarkable Life of NASA's Visionary Leader George M. Low
Richard Jurek
Foreword by Gerald D. Griffin

To order or obtain more information on these or other University of Nebraska Press titles, visit nebraskapress.unl.edu.